AF389147

Durum Wheat Breeding and Genetics

Durum Wheat Breeding and Genetics

Editors

Pasquale De Vita
Francesca Taranto

MDPI • Basel • Beijing • Wuhan • Barcelona • Belgrade • Manchester • Tokyo • Cluj • Tianjin

Editors
Pasquale De Vita
Research Centre for Cereal and Industrial Crops (CREA-CI)
Italy

Francesca Taranto
Research Centre for Cereal and Industrial Crops (CREA-CI)
Italy

Editorial Office
MDPI
St. Alban-Anlage 66
4052 Basel, Switzerland

This is a reprint of articles from the Special Issue published online in the open access journal *Agronomy* (ISSN 2073-4395) (available at: https://www.mdpi.com/journal/agronomy/special_issues/durum_wheat_breeding_genetics).

For citation purposes, cite each article independently as indicated on the article page online and as indicated below:

LastName, A.A.; LastName, B.B.; LastName, C.C. Article Title. *Journal Name* **Year**, *Article Number*, Page Range.

ISBN 978-3-03943-102-1 (Hbk)
ISBN 978-3-03943-103-8 (PDF)

Cover image courtesy of Pasquale De Vita.

Contents

About the Editors

Pasquale De Vita (Ph.D.) is a Senior researcher at CREA Research Centre for Cereal and Industrial Crops, Foggia, Italy. His main interest is focused on durum wheat breeding and genetics, including QTL mapping, marker-assisted selection, phenotypic and molecular analysis of germplasm collections for breeding purposes. His current research includes the evaluation of Triticum ssp. and their genetic diversity through a multidisciplinary approach, with the objective of identifying new sources of useful traits/genes to be exploited in breeding programs and varietal improvement. Dr. De Vita has authored and co-authored more than 80 peer-reviewed journal articles, including review papers and book chapters. He also holds two industrial patents and has developed six durum wheat varieties in collaboration with seed companies.

Francesca Taranto (Ph.D.) is a Researcher at the National Research Council of Italy, Institute of Biosciences and Bioresources (CNR-IBBR), Portici (Napoli), Italy. In 2013, she defended her Ph.D. thesis in "Plant breeding and pathology of agricultural and forestry plants" at the University of Bari, Italy. Her research activity is focused on marker-assisted breeding and genetics of wheat, legumes, and cultivated trees (mainly olive and grapevine). She employs genetics, molecular biology, genomics, and bioinformatics techniques to investigate population genetics, biodiversity, and gene flow and identify key genes involved in the expression of important agronomic traits. Since January 2018, she is a Review Editor in Bioinformatics and Computational Biology for the journal Frontiers in Genetics and a Review Editor for Plant Breeding, a specialty section of Frontiers in Plant Science. Dr. Taranto has authored and co-authored more than 30 peer-reviewed journal articles, including review papers and book chapters.

Preface to "Durum Wheat Breeding and Genetics"

Durum wheat is grown primarily in the Mediterranean basin, with a total production that varies significantly every year due to unpredictable climatic conditions. The frequency and the intensity of extreme weather events are concomitant with changes in global climate, making the selection of new durum wheat varieties difficult. The Special Issue Book "Durum Wheat Breeding and Genetics" is based on scientific papers addressing major concerns related to the genetic improvement of durum wheat. Seven chapters including two review papers provide an update on the progress in the genetic improvement of durum wheat, suggesting traits and strategies to maintain productivity and high qualitative standards, despite increasing water scarcity and higher temperatures. It is necessary to exploit the best modern technologies and the entire methodological arsenal currently available to prevent the stagnation of durum wheat production. Understanding the genetic bases of variation for the most important agronomic traits and identifying allelic variants associated with tolerance to abiotic stresses of durum wheat are a priority. To this end, it is necessary to explore the genetic variability existing within durum wheat species, including landraces and traditional varieties. This special issue offers new breeding opportunities for selecting modern varieties adapted to climate change and expanding the durum wheat production.

Pasquale De Vita, Francesca Taranto
Editors

Review

Durum Wheat Breeding in the Mediterranean Region: Current Status and Future Prospects

Ioannis N. Xynias [1,*,†], Ioannis Mylonas [2,†], Evangelos G. Korpetis [2], Elissavet Ninou [3], Aphrodite Tsaballa [2], Ilias D. Avdikos [4] and Athanasios G. Mavromatis [4]

[1] School of Agricultural Sciences, University of Western Macedonia, Terma Kontopoulou, 53100 Florina, Greece
[2] Institute of Plant Breeding and Genetic Resources, Hellenic Agricultural Organization–"Demeter", 57001 Thessaloniki, Greece; ioanmylonas@yahoo.com (I.M.); korpetis@yahoo.gr (E.G.K.); tsaballa80@gmail.com (A.T.)
[3] Institute of Industrial and Fodder Crops, Hellenic Agricultural Organization Demeter–"Demeter", 41335 Larissa, Greece; lisaninou@gmail.com
[4] Laboratory of Genetics and Plant Breeding, School of Agriculture, Aristotle University of Thessaloniki, 54124 Thessaloniki, Greece; avdikos.elias@gmail.com (I.D.A.); amavromat@agro.auth.gr (A.G.M.)
*   Correspondence: ixynias@teiwm.gr or ioannis_xynias@hotmail.com
†   These authors contributed equally to this work.

Received: 10 January 2020; Accepted: 19 March 2020; Published: 21 March 2020

Abstract: This brief historical review focuses on durum wheat domestication and breeding in the Mediterranean region. Important milestones in durum wheat breeding programs across the countries of the Mediterranean basin before and after the Green Revolution are discussed. Additionally, the main achievements of the classical breeding methodology are presented using a comparison of old and new cultivars. Furthermore, current breeding goals and challenges are analyzed. An overview of classical breeding methods in combination with current molecular techniques and tools for cultivar development is presented. Important issues of seed quality are outlined, focusing on protein and characteristics that affect human health and are connected with the consumption of wheat end-products.

Keywords: Mediterranean basin; durum wheat; breeding; MAS; biochemical markers; quality

---

1. Introduction

Durum wheat (*Triticum turgidum* subsp. *durum* (Desf.) Husnot) is one of the most essential cereal species and is cultivated worldwide over almost 17 million ha, with a global production of 38.1 million tonnes in 2019 [1]. The largest producer is the European Union, with 9 million tonnes in 2018, followed by Canada, Turkey, United States, Algeria, Mexico, Kazakhstan, Syria, and India [2–6]. Durum wheat production and cultivation areas are concentrated in the Mediterranean. Moreover, the countries of the Mediterranean basin are the largest importers and the largest consumers of durum wheat products (flour, pasta, and semolina). Among European Union (EU) countries, Italy is considered the leader of durum wheat production, with an average production of 4.26 million tonnes in the last decade (1.28 million ha growing area), followed by France with 1.89 million tonnes (0.37 million ha), Greece with 1.07 million tonnes (0.37 million ha), and Spain with 0.98 million tonnes (0.38 million ha) (Table 1) [2]. Since durum wheat is mainly grown under rain-fed conditions in the Mediterranean basin, its productivity is profoundly affected by rainfall and biotic (pests and diseases) and abiotic (drought, sunlight, cold, and salinity) stresses.

Table 1. The world's leading durum wheat producing countries (2010–2019) [2–4].

Country	Average Production (Millions of Tonnes)
Canada	5.2
Italy	4.3
Turkey	3.7
USA	2.3
Kazakhstan	2.2
Syria	2.2
Algeria	2.2
France	1.9
Morocco	1.8
Greece	1.1
Spain	1.0
Tunisia	1.0

2. A historical Review of Durum Wheat Breeding

2.1. Prehistory and Early History

Wheat cultivation and human civilization have evolved together for at least 10,000 years since humans first attempted to produce food during the "Neolithic Revolution" [7]. The first step in the improvement of durum wheat involved the domestication of its wild progenitors [8] in the hilly area of southwest Asia at the Tigris and Euphrates basin (ancient Mesopotamia) and in the mountains of Iran, Turkey, Syria, and Jordan, in the area often referred to as the "Fertile Crescent" [9,10].

According to Shewry [7], an early and primitive form of plant breeding was carried out with the first selections from wild populations of *T. turgidum* subsp. *dicoccoides* (Körn. ex Asch. and Graebner) Thell. based primarily on yield, but also considering other genetic characteristics; mainly non-brittle rachis and free-threshing naked kernels. It has been proven that the first trait is controlled by two major genes, *brittle rachis 2 (Br-A2)* and *brittle rachis 3 (Br-A3)*, on the short arms of chromosomes 3A and 3B, respectively [11], while the free kernels originated from a dominant mutation at the Q locus [7,12].

The cultivated emmer (*T. turgidum* subsp. *dicoccum* (Schrank ex Schübler) Thell.) was the first dominant wheat in Asia, Africa, and Europe in the early years of agriculture, in the seventh millennium BC. Emmer grains were found in the tombs of the Egyptian Pyramids and were widespread in ancient Babylon and Central Europe [13]. Seeds of emmer were discovered in archaeological finds dating back to between 7500 and 6700 BC [10]. It remained a dominant cereal until the first millennium BC when it was replaced by free-threshing durum wheat [14]. Its grains have been found in eight archaeological sites of the Iron Age, dating from the end of the 2nd millennium BC to the end of the 4th century BC in Northern Greece [15]. The first agricultural book on wheat was written using cuneiform on a clay tablet around 1700 BC and was discovered in Israel in 1950 [13].

The written history of wheat science began 2500 years ago when the Greek botanist Theophrastus (371–287 BC) wrote the study "Enquiry into Plants". Later, the famous Roman writers Columella, Varrone, Virgil, and Pliny the Elder described wheat as the primary agricultural food source in the Mediterranean region [16].

2.2. Modern History

The modern history of durum wheat breeding in the Mediterranean region began in the early 20th century and was affected by the global evolution of agricultural science. A high yield, good end-use traits, and resistance to abiotic and biotic stresses have always been targets for wheat breeders. The initial approach in durum wheat breeding focused on the exploitation of local genetic resources. Later, the Green Revolution resulted in the release of short, high-yielding durum wheat cultivars from International research institutions, which are used extensively in crosses in almost all national breeding programs.

2.3. The Early Period, before the Green Revolution

At the dawn of the 20th century, the first systematic breeding efforts were made by public research institutes in the countries across the Mediterranean basin that pioneered plant breeding at that time. Initially, landraces were used as the starting materials in breeding programs (Table 2). Later, systematic breeding schemes were gradually applied using parent cultivars with different useful agronomic characteristics, such as high yield, good quality, and resistance to a range of disease.

Table 2. The main Mediterranean wheat landraces.

Country	Reference
Italy	
Biancuccia, Bivona, Castiglione, Ciciredda, Cotrone, Duro Lucano, Farro Lungo, Gioia, Regina, Ruscia, Sammartinara, Timilia	[17]
Greece	
Roussias, Asprostachys, Tsipoura, Asprostaro, Diminitis, Trimini, Gremmenia, Kondouzi, Mavragani, Zochovis, Deves, Rovaki, Arnaouti, Kokkinostaro, Floritsa, Mavrostaro	[18,19]
Turkey	
Uveyik, Sahman, Bagacak, Sorgul, Havrani, Menceki, Iskenderi, Kocabugday, Cambudayi, Kibris bugdayi, Amik, Akbasak, Karabasak, Karakilcik, Kunduru, Sari Bursa, Sari Basak	[20,21]
Algeria	
Tuzelle, Mahon, Biskri, Bidi 17	[22]
Morocco	
ML 19, ML 21, ML 22, ML 23, ML 24, ML 26, ML 28, ML 48, ML 49, ML 32, ML 33, ML 34, ML 35, ML 36, ML 37, ML 38, ML 39, ML 41, ML 42, ML 43, ML 44, ML 45	[23]
Tunisia	
Hamira, Richi, Mahmoudi, Jenah Khotifa	[24]

In Italy, breeding was initially done through public research institutes, universities, and departments from the Ministry of Agriculture [25]; later, private companies such as the Società Italiana Sementi and Società Produttori Sementi Bologna became involved [26]. The pioneer in the modern durum wheat breeding was the Italian Nazareno Strampelli (1866–1942). Strampelli began his efforts with selections of local landraces from southern Italy, the Italian islands, and the Mediterranean region. In 1923, Strampelli released the cultivar "Senatore Capelli", which he had selected in 1915 from the local North African landrace "Jean Retifah". This cultivar was a landmark for the cultivation of durum wheat in Italy, as it covered 60% of Italy's durum wheat acreage for many decades, but also throughout the Mediterranean, where it has been widely used for crosses [16,27]. Casale, an Italian breeder, crossed cv. "S. Capelli" with Palestinian cv. "Eiti 6" in the 1940s and released the cultivar "Capeiti 8" which replaced "S. Capelli" in durum wheat cultivation in 56% of acreages due to its higher yield, although its seed quality was low. Another important breeder, Forlani, saw the possibility of improving durum wheat using interspecific crosses, particularly to introduce resistance to diseases [16,28].

In Greece, modern wheat breeding was started in 1923 by Ioannis (Juan) Papadakis (1903–1997), who founded the Institute of Plant Breeding, the first Research Center in the Balkan Peninsula. Papadakis introduced the new methods of the era, such as selection and crossbreeding, and conducted experiments in different locations using different controls and replications to evaluate the wheat's genetic material. Papadakis started by inserting selections into local breads and durum wheat landraces. In 1924, he made the first crosses by selecting parents from both local and foreign landraces (Table 3) [29]. In 1929, Papadakis recorded and described in detail the genetic material of the wheat found in Greece, according to Professor Percival's classification [30]. This study included 47 local cultivars or landraces of durum wheat cultivated in Greece [18]. The first durum wheat cultivars were released in 1932 [31].

The local durum wheat landraces were cultivated until 1930 were then replaced gradually by the new improved cultivars. The crosses of the plant breeding program of the Institute of Plant Breeding were produced in 1946, resulting in the cultivars "Methoni" and "Electra", which had better quality and featured earlier maturation by one week compared to "Lemnos". After 30 years, the new improved cultivars covered 60% of the durum wheat cultivated area in Greece, with the improved cv. "Lemnos" occupying 42% of the area [32].

Table 3. The main Mediterranean cultivars released before the Green Revolution.

Country	Name	Year of Release	Pedigree, Cross Name, Origin	Agronomic Characteristics	Reference
Italy	Senatore Capelli	1923	Selection from landrace "Jean Retifah"	Waxy, tall, rather late ripening, good quality	[16,17,27]
	Capeiti 8 (=Patrizio 6)	1955	S. Capelli × Eiti 6	Better yield and poorer quality than S. Capelli	[16,17,28,33]
	Sincape 9		Sinai × S. Capelli	Agronomical characteristics superior to S. Capelli	[16,28,33]
	Appulo	1964	(S. Cappelli × Grifoni 235) × Capeiti 8	Optimal grain quality and yield	[16,28,33]
	ISA-1	1971	Patrizio × Sassari 0130	Very early	[28]
Greece	Lemnos	1932	Selection from Landrace Akbasak	High yield, good quality for bread	[31]
	Methoni	1954	Lemnos × 7-B-1231	Better quality and earlier maturing than Lemnos	[32]
	Electra	1957	S. Capelli × [Lemnos × (Florence × Arditto) × Sinai2]	Better quality and earlier maturing than Lemnos	[32]
Spain	Andalucia 344	Before 1940	Selection from Manchón de Alcalá la Real		[34,35]
	Hibrido D	Before 1940	S. Capelli/Colorado dc Jerez		[34]
	Ledesma	Before 1940	S. Capelli/Rubio dc Belalázar		[34]
	Bidi 17	1950	Selection from Bidi or Blé Gounod, from Algeria	Tall, large grains and a weak yellow color index.	[34]
France	Bidi 17	1950	Selection from Algerian population "Oued Zenatti"	Tall, large grains and a weak yellow color index.	[26,36]
	Montferrier, Mandon			Better yield, very sensitive to leaf rust	[28,36]
	Agathe			Better yield, moderately sensitive to leaf rust	[36]
Turkey	Kunduru 1149	1967	Selection from Landrace	Tall, drought tolerant, good winter hardiness	[20,21,37]
	Berkmen 469	1970	Selection from Landrace	Tall, drought tolerant, good winter hardiness	[21,37]
Morocco	Oued.Zenati	1949	Selection from landraces	Tall	[23,38]
	Zeramek	1949	Selection from landraces		[23,38]
	Kyperounda	1956	Selection from landraces	Tall	[23,38]
Tunisia	Chili 931	1953	France		[24]
	Kyperounda	1954	Cyprus		[24]
	INRAT 69	1970	Mahamoudi981/Kyperounda		[24,39]

At this initial stage of breeding, the classification of *Triticum* species was also the subject of extensive study by Spanish researchers. In the early 19th century, Spanish botanists Clemente and Lagasca compiled the "Ceres hispanica", the first herbarium for *Triticum* species. Intensive work on the collection and conservation of durum wheat types in the Iberian Peninsula began in 1930 by Arana and was extended from 1950 to 1956 by Tellez, Prieto, and Garcia-Pozuelo. The first attempt to improve durum wheat in Spain was performed at the Agricultural Research Center of Jerez de la Frontera by Juan Bautista Camacho [40]. Based on the work that had been done in the previous years, selections were made in local durum wheat populations, and crossing programs were carried out, which resulted in the release of important cultivars, such as "Ledesma", "Andalucia 344", and "Hibrido D", which, in 1963, covered 12%, 10%, and 6% of the durum wheat cropping area in Spain, respectively [34].

The characteristic resistance of local landraces to three types of rust has been studied at the Instituto Nacional de Investigaciones Agrarias since 1954 [40].

Concerning France, until the 1960s, production and research programs have mainly been conducted in Algeria, Morocco, and Tunisia. In 1930, breeding work was conducted in Algeria by Ducellier, who identified 29 local cultivars or populations of durum wheat. Wheat research was performed and coordinated by the Institut National de la Recherche Agronomique (INRA) in Montpellier until Algeria's independence in 1962, and then from The Algerian Center for Agronomic, Scientific, and Economic Research (CARASE), or INRA, in Algeria. In the early 1970s, the production of durum wheat increased drastically due to the release of the cultivars "Bidi 17", "Oued Zenati 368", "Hedba 3", and "Mohamed Ben Bachir", which resulted from selections among local landraces [22]. Accordingly, before the 1960s, breeding efforts in Morocco were focused on collecting local durum wheat landraces, from which, after the selection programs, the cultivars "Oued Zenati", "Zeramek", and "Kyperounda" were released (from 1949 to 1956). The latter cultivars were more productive than the local landraces, but their quality was inferior [23]. A similar approach was recorded in Tunisia, where local durum wheat landraces were mainly cultivated in the country until the 1940s. These were followed by cultivars produced from selections within the aforementioned landraces [24].

The Ministry of Agriculture started a breeding program in Libya in the 1950s. Durum wheat landraces, such as "Jlail-Dib" and "Hmira", were used to develop improved varieties. During the period of 1962–1965, nine durum wheat cultivars were collected and characterized [41]. Simultaneously, breeding work for durum wheat took place in Egypt, another important North African country. In 1914, Egyptian breeders released two cultivars with resistance to high temperatures, "Dakar 49" and "Dakar 52", which were selected from local durum wheat landraces. In the 1920s, these breeders released the cultivars "Hindi D", "Hindi 62", and "Tosson" using genetic material imported from India. Since 1942, this breeding program has focused on creating cultivars with resistance to the three aforementioned types of rust [42].

In Turkey, initial wheat research began in 1925 with the establishment of the first "Seed Improvement Station" in Eskisehir. The early durum wheat cultivars that were released resulted from selections within the local durum wheat populations and included "Kunduru 414/44", "Sari Bursa 7113", "Kunduru 1149", "Berkmen 496", "Akbasak 073/44", and "Karakilcik 1133" [20,43]. Finally, in Israel, the onset of genetic improvement in durum wheat was based on selections within local landraces. Local wheat landraces, mainly durum wheat that was previously cultivated in Palestine, were described as early as the 1920s. In 1948, Kostrinsky compiled a descriptive list of cultivars and landraces [44].

2.4. Breeding Work in the Mediterranean during and after Green Revolution

A significant increase in yield was achieved in many national breeding programs through the second half of the 20th century [45]. The key to the dramatic increase in wheat yields during the 'Green Revolution' in the 1960s was the introduction of semi-dwarf genes into bread wheat, which resulted in the replacement of tall cultivars with semi-dwarf cultivars characterized by their responsiveness to inputs (e.g., fertilizers) and resistance to lodging. The primary donor of these genes was the semi-dwarf cultivar "Daruma" from Japan. Initially, the cultivar "Norin 10" was developed as a result of diallelic crosses between "Daruma" and some American cultivars. Norman Borlaug, at the Centro Internacional de Mejoramiento de Maiz y Trigo (CIMMYT), used the cross "Norin 10" x "Brevor 14" for the development of initial genetic material that was first shipped to Latin America and Southeast Asia, where it was rapidly adopted for cultivation with spectacular increases in yield [46]. The cultivar "Norin 10" has also been widely used in breeding programs around the world for the introduction of semi-dwarf genes in durum wheat, which has been recorded since 1956 [47]. This was an important milestone that affected durum wheat breeding efforts worldwide. The CIMMYT durum wheat germplasm continues to play an essential role in increasing the production and genetic gain of national agricultural research systems for developing countries. Indeed, 77% of the cultivars released in developing countries in 1991–1997 originated from crosses between CIMMYT and indigenous genotypes. Moreover, 19% of these included at least one parent from the CIMMYT collection [48].

Another critical event in the evolution of plant breeding was the foundation of the International Center for Agricultural Research in the Dry Areas (ICARDA) in 1977 in Aleppo, Syria. Initially, this center was intended to control the CIMMYT's segregated wheat genetic material and pure lines. Since 1981, the ICARDA Improvement Program has worked on crosses, and, in 2003, this program became independent [49,50]. The first successful cultivar of durum wheat resulting from ICARDA's breeding program was "Waha" [syn. Cham1, Plc/Ruff//Gta/Rtte], which was enlisted in Algeria's National Variety Catalogue in 1984. Another successful cultivar was "Om Rabi 1" [syn. Cham 5, Jori/Haurani], which was released in Morocco in 1989. As of 2018, 130 cultivars that originated from the genetic material of ICARDA's breeding program have been released in 22 countries (Table 4) [50].

A comparison between old and new durum wheat cultivars bred in different periods is useful to understand which of the yield components and other associated traits contribute to the increased productivity of new cultivars. In ICARDA, the impact of 35 years (1977–2011) of public durum wheat breeding resulted in a 0.7 genetic gain per year, mostly based on earlier flowering and an increase in the spike density [50]. In Canada and Italy, durum wheat yield improvement is based on an increased number of grains [51,52]. Similarly, new high yield CIMMYT cultivars increased their numbers of grains per m^2 as a result of the increased number of spikes per m^2 and the grains per spike [48,53]. Royo et al. [54] studied the genetic changes in durum wheat yield components and their associated traits by comparing 24 old (<1945), intermediate (1950–1985), and modern (1988–2000) Italian and Spanish cultivars. As a result, it was found that the number of grains per m^2 increased by 39% and 55% in Italian and Spanish cultivars, respectively. This increase involves a 20% increase of plants per m^2, a 29% increase of spikes per plant, and a 51% increase of grains per spike. The mean rate of increase in the yield per plant was 0.41% per year, 0.11% per year for plant density, 0.55% per year for the number of grains per m^2, 0.48% per year for the harvest index, and 0.81% per year for the plant height [54]. The number of spikelets per spike did not change, so the increased numbers of grains per m^2 was due to the higher grain set in the modern cultivars. Similar results were found in a study on durum wheat cultivars released in Italy during the 20th century, where the genetic gain was mostly based on a higher kernel number per m^2 and spikes per m^2 [52].

Apart from yield, durum wheat breeding programs significantly affect grain quality [52,55]. The evaluation of durum wheat cultivars released during different breeding eras shows that genetic improvement reduces grain protein concentration as a result of improved yields, but without affecting pasta cooking quality [53]. The lower grain protein percentage of modern cultivars is based on the dilution effect caused by their heavier grains or increased amount of carbohydrates [55,56]. However, modern cultivars have increased gluten index, showing an improvement in pasta-making quality [32,52,55,57].

Table 4. The main Mediterranean cultivars released after the Green Revolution.

Country	Name	Year of Release	Pedigree, Cross Name Origin	Agronomic Characteristics	Reference
Italy	Trinakria	1973	(CpB144=Castelfusano) × {[(Yaktana54-Norin10-Brevor) Capelli-63-4] Tehucan}	Good quality	[16,26]
	Creso	1974		Good quality, resistance to *Fusarium graminearum* and brown rust	[16,17,33]
	Simeto	1988	Capeiti 8 × Valnova	High yield, low tillering, excellent adaptability	[17,33]
	Svevo	1996	Sel. CIMMYT × Zenit sib	High quality for pasta	[16,26]
Greece	Mexicali-81	1981	Selection from Mexicali 75	High yield	[58]
	Selas	1982	Selection from Stork "S"	Good grain quality	[58]
	Elpida	2010	Sifnos × Mexicali-81	High quality and yield	[58,59]
	Thraki	2014	Simeto × Mexicali-81	High quality and yield	[58,60]

Table 4. *Cont.*

Country	Name	Year of Release	Pedigree, Cross Name Origin	Agronomic Characteristics	Reference
Spain	Claudio	1999	(Sel. Cimmyt × Durango) × (IS193B×Grazia)		[61]
	Mexa	1980	GERARDO-VZ-469/3/JORI(SIB)// ND-61-130/LEEDS		[35,61]
	Vitron	1983	TURCHIA-77/3/JORI-69(SIB)/ (SIB)ANHINGA//(SIB)FLAMINGO		[35,61]
	Cocorit	1978	CIMMYT		[35]
France	Durtal	1972	*Triticum durum × T. aestivum*	High yield, short, good earliness, poor quality	[26,28,36]
Turkey	Dicle 74	1975	RAE/4×TC//STW63/3/AA"S" =Cocorit	Spring wheat	[20,37]
	Gediz 75	1976	LD357E/TC2//JO"S"	Spring wheat	[20,37]
	Cakmak 79	1979	UVY162/61.130	Winter wheat, good quality	[20,21,37]
	Kiziltan91	1991	UVY162/61.130	Winter wheat, good quality	[20,21,37]
	Altintac 95	1995	KND//68111/WARD	Irrigated winter wheat	[21,37]
	Selcuklu 97	1997	073/44×2/OVI/3/DF-72//61-130/ UVEYYK 162	Irrigated winter wheat	[21,37]
	Yilmaz98	1998		Irrigated winter wheat	[20,21,37]
	Ankara98	1998		Irrigated winter wheat	[20,21,37]
Algeria	Waha	1984	syn. Cham1, Plc/Ruff//Gta/Rtte	ICARDA genetic material	[50]
Morocco	Cocorit			Yield potential, wide adaptation, low quality	[62]
	Jori			Yield potential, wide adaptation, grain quality	[62]
	Haj-Mouline			Yield potential, wide adaptation, low quality	[62]
	Oum Rabia	1988	INRA 1718, Sel in " Cyprus 3"	High yield potential, better grain quality	[23,38,62]
	Karim	1985	Bittern 'S' or sel in « JO'S'.AA":S'//FG'S' »	High yield potential, better grain quality	[23,38,62]
Tunisia	Karim	1982	(Jori"S"/Anhinga"S"//Flamingo"S") CIMMYT		[24,39]
	Khiar	1992	Chen"S"/Altar 84, CIMMYT		[24]
	Om Rabia	1987	Jori C69/Haurani, ICARDA		[24]
	Nasr	1990	GoVZ512/Cit//Ruff/Fg/3/Pin/Gre//Trob), ICARDA		[24]
	Maali	2003	CMH80A.1016/4/TTURA/CMH74A370/ CMH77.774/3/ YAV79/5/Rassak/6/DACK"S"/YEL3"S"// Khiar, INRAT		[24]
	Salim	2010	ALTAR 84/FD8419-126-1-2/Razzak/3/Krf/ Baladia Hamra, INRAT		[24]

In tandem with the evolution of agricultural science occurring worldwide during the same time, in 1956 in Italy, Gian Tommaso Scarascia Mugnozza and Francesco D'Amato embarked on a pioneering durum wheat breeding program that included both fundamental genetic studies and applied mutation breeding and led to 22 registered varieties, six of which resulted from a direct selection of induced mutants [16,63]. An important cultivar that emerged from this program in 1974 was "Creso". Considerable work has also been done at Italian universities, such as at Palermo University (cv. "Trinakria" in 1973 by G.P. Ballatore), at Sassari and Naples (cvs. "Maristella", "Ichnusa" by R. Barbieri), and at Bari University [cvs. "Messapia", "Salentino", "Salizia" by G.T. Scarascia-Mugnozza, A. Blanco and coworkers). J. Vallega and G. Zitelli, at the Experimental Institute for Cereal Research in Rome, used N. Borlaug's selected genetic material in their crosses in their attempt to produce high-yielding cultivars resistant to lodging and diseases; this material had semi-dwarf genes of "Norin 10" [16]. In 1988, Calcagno released the successful cultivar "Simeto", which was bred at the Experimental Station for wheat at Caltagirone in Sicily. This cultivar was high yielding, low tillering, and had excellent adaptability across different environments. Thus, it has been widely cultivated in all the countries of the Mediterranean basin up to present with a large acreage [26] and is often used as a parent in crosses. In the private sector, seed companies have released many notable cultivars, including "Duilio" (Società Italiana Sementi, S.I.S.) and "Svevo" (Società Produttori Sementi), which are cultivated in several important durum wheat productive countries offering high-quality pasta products.

In Greece in the late 1960s, E. Skorda induced artificial mutations with gamma rays and thermal neutrons to increase genetic variability [64]. In the early 1980s, the breeder's team from the Cereal Institute in Thessaloniki released cultivars that were bred from intra-cultivar selections of genetic material from the Mediterranean basin, including "Mexicali 81", "Kallithea", "Athos", and "Selas". Furthermore, new cultivars selected from CIMMYT segregating germplasm were made available to the farmers. Among the aforementioned released cultivars, "Selas" had an excellent grain quality and was used until 2015 by the pasta industry in Greece. The newest durum wheat cultivars released by the breeders from the Cereal Institute originated either from crossbreeding among different cultivars or from selections among CIMMYT segregating material included in the Greek National and the Common Catalogue of Cultivated Plants [58,60].

At the beginning of the 1960s in France, Pierre Grignac began the first durum wheat breeding program, in which the first crossings among Mediterranean landraces were developed at INRA in Montpellier. In the middle of the same decade, cultivars with good qualities, such as a good yellow-amber color, were imported from North Dakota for cultivation in northern France, and constituted a second genetic pool for the program [26]. Afterwards, Grignac used interspecific crosses with bread wheat to select new lines with improved characteristics [28,36].

In Algeria, the Field Crops Development Institute (IDGC) was founded in 1974. This institute was responsible for durum wheat breeding and has collaborated actively with CIMMYT and ICARDA since 1980. As a result, more than 60 modern durum wheat cultivars have been released, and the peak yields have been obtained with the cultivars "Hoggar" and "Sahel" [22]. In Tunisia, modern cultivars developed from CIMMYT and ICARDA genotypes prevailed during the 1970s and were replaced in the 1980s by more recent cultivars developed at INRAT [24].

In 1970, a result of the collaboration with CIMMYT and, later, with ICARDA was the introduction of foreign durum wheat germplasm into Morocco in an attempt to develop new high-yielding, early-maturing cultivars. The outcome of this program was the release of the cultivars "Marzak", "Karim", "Sebou", and others after 1984 [23]. However, this new germplasm was inferior in quality.

Finally, in Turkey, the agreement between the Turkish government and the Rockefeller Foundation in 1967 resulted in the release of semi-dwarf, high-yielding, and disease-resistant cultivars [20,65].

3. Application of Classical Methodologies of Breeding

3.1. Classical Approaches, New Perspectives, and Tools for Wheat Breeding

The main targets of a durum wheat breeding program established in the Mediterranean basin, where its cultivation is very well adapted, may focus on: (1) grain yield improvement; (2) yield stability and a better understanding of genotype × environment interactions (G × E) and adaptation mechanisms; (3) responsiveness to inputs and the use efficiency of recourses in different cultivation systems; (4) resistance to biotic stresses (pests and diseases), and tolerance to abiotic stresses (drought, salinity, etc.); or (5) improving grain quality.

The above parameters are taken into account by durum wheat breeders in the design of a breeding program for cultivar development. They use common classical breeding methods, such as pedigree, bulk, single-seed, backcross method, pure line selection, and recurrent selection, to develop cultivars with the desired characters mentioned above. These methods can be applied with some alterations to increase efficiency and reduce the duration of the breeding program. As an example, backcross (BC) is a very common method mainly used in durum wheat for the transfer of traits controlled by only one or a few genes, such as resistance to diseases or quality parameters, from one donor parent to an elite line (recurrent parent) [66,67]. Molecular markers can significantly shorten the time needed, compared to the conventional backcross method, to identify the desired plants that have the target gene/genes and apply negative selection for the donor genome to ensure the maximum recovery of a recurrent-parent genome [68]. There are successful results in durum wheat breeding programs regarding the increased disease resistance or quality characteristics. Marker-assisted backcrossing (MABC) through simple sequence repeat (SSR) markers was shown to improve grain protein content in a wheat cultivar [69]; this method can assist the simultaneous selection of multiple stripe rust resistance genes and help avoid escapees during the selection process. In a previous study, the successful targeting of the gene transfer and reconstitution of the genome were completed in a period of four crop cycles, proving the practical application of MABC in developing high grain protein lines in the background of any popular cultivar [69]. In the single-seed descent method (SSD), only one seed from each F_2 selected plant is kept and bulked with all the others to produce an F_3 population. The same procedure is repeated until the F_5–F_6 generation. Two or more generations grow per year in the greenhouse, in winter nurseries, or in a growth chamber. The selection of lines takes place in the F_6 lines, which have increased homozygosity, retaining a large part of the extensive genetic variability from the F_2 generation. The single-seed descent method is considered a tool to exploit durum wheat genetic resources [70]. In the bulk breeding method, some plants are selected from the F_2 population, and their seeds are bulked to form the F_3 population. The same procedure is repeated until the F_5 generation, where the evaluation of lines begins until the F_{10}–F_{11} generation. This is an easy method to apply, thus saving breeders time and effort. Also, this method can increase the frequency of desirable genotypes in a population, but is not very effective for traits with low heritability. However, a modification of this method could be useful in wheat breeding when applied under salinity stress conditions [71]. In this case, using molecular markers through a bulk segregant analysis (BSA) will increase the effectiveness and shorten the needed time for all these processes [72].

Finally, the pedigree breeding method (and its modifications) is the most common method used in breeding programs for the release of durum wheat cultivars. Generally, pedigree breeding includes phenotypic selection in the early generations (until F_{3-4}), and the normal yield test begins in the generations with increased homozygosity (~F_5). The selection for yield during early generations was not very effective when the evaluation was done in normal plant density fields [73–76]; however, it was found to be effective when the evaluation was applied in low plant density fields and the experimental unit involved a single plant, as in the honeycomb methodology [77–80]. Finally, the selected lines from the experiments with replications will be evaluated in large plots (drill strips) over ~2 years and ~5–10 locations to determine yield, stability across locations, maturity, plant height, semolina, pasta and important quality characteristics for cooking [81]. Following a classical approach, a period of 9–12 years is needed from the beginning to the end of a breeding program, but this time could be reduced significantly by using the Marker Assisted Selection (MAS) procedure [82].

It has been recorded that the effectiveness of breeding on durum wheat's yield potential has been remarkable in Spain, Morocco, Turkey, and Italy [20,35,62,83]. The improvement of grain yield (GY) during the past decade has been attributed mostly to the increased number of grains m^{-2} and to the increased number of spikes m^{-2} [83]. Thus, further improvements in these characteristics might improve yield [83]. Important increases in yield were also achieved by increasing the harvest index, which has almost doubled since the beginning of the 20th century [83,84]. Moreover, a further increase in yield was achieved through an increase in biomass and a subsequent increase in yield, which explains the stability of the harvest index over the last three decades [84]. Similarly, an increase in biomass could result in an increase in yield in the future. Other characteristics that contributed to the increased yield in the 20th century include a reduction in the heading date and physiological maturity and an increase in the grain filling period [83]. Similar changes in these characteristics in the future could result in an increase in yield, based on the results of the reduction of the effects of drought and heat stress. Moreover, an increased yield could be achieved through the release of cultivars with higher water use efficiency [85].

Apart from the selection for increased yield potential, further improvements could be achieved through selection for increased tolerance to abiotic and biotic stresses. Moreover, the importance of traits that allow a plant to escape terminal drought and avoid critical stages of seed development (anthesis and seed filling), such as early vigor and an early heading date, has been well recognized [86,87]. All the above classical breeding methodologies have succeeded in making considerable progress in the yield and quality of durum wheat in the Mediterranean basin [24,57,88]. The main problem for the classical methods is that they are time-consuming, and phenotyping procedures are costly. In an effort to aid classical breeding methods, molecular genetics and associated technologies have been developed, and they offer important tools for plant breeders.

The parental selection of wheat lines can be based on phenotypic characterization and biochemical and DNA markers, which can estimate genetic variability even among phenotypically similar genotypes, as identified in several studies undertaken in Mediterranean countries [89–93]. By employing molecular markers in parental selection, the genetic diversity of wild and cultivated wheat can be exploited [94–98]. As an example, SSR markers were proven to be effective in the selection of genetically diverse genotypes with phenotypical similarities [90]. A combination of molecular markers and pedigree data could help in the exploitation of genetic diversity [91,99] and the selection of progenies, significantly increasing the efficiency and precision of plant breeding programs. Molecular markers supply various advantages over morphological markers in the linkage mapping of important agronomic traits. They are also unlimited in number, highly polymorphic, and can be used at any developmental stage without any environmental interference. Molecular markers can increase the precision and speed of selection in a durum wheat program though: (a) selection in the early stages or a simultaneous selection of multiple traits or traits that are difficult or expensive to evaluate; (b) the targeted introgression of useful genes in wide crosses; and (c) accelerated backcrossing. MAS or molecular breeding offers an opportunity to accelerate classical breeding approaches. MAS requires the establishment of a correlation between a desired trait, such as disease resistance, and molecular marker(s); this can be obtained by phenotyping a genetic mapping population followed by a quantitative trait locus (QTL) analysis [81]. For this purpose, several markers that are known to be associated with QTL/genes for some major economic traits are being deployed for MAS in wheat breeding programs. Several examples of the successful use of MAS are now available for wheat, and more examples will become available in the future [100–102]. Furthermore, molecular allele mining can help in broadening the reduced genetic diversity of cultivated wheat through the identification of allelic variation and the isolation of new rare alleles capable of improving tolerance to abiotic and biotic stresses [103,104]. According to Sehgal et al. [104], by using new technologies, unexploited genetic variation can further improve the drought and heat stress tolerance of the elite wheat pool and enrich it with novel drought and heat tolerance genes. This will contribute to achieve adaptability of the released cultivars to high temperature and drought that is for the most important emerging problems emerging in the Mediterranean due to climate change.

3.2. Participatory Plant Breeding

Today, it is recognized that agricultural production requires the adoption of environmentally friendly solutions, the preservation of crop biodiversity, and the release of varieties suitable for low input environments to set new goals for wheat breeding that align with the real needs of farmers and the market that are imprinted in the Mediterranean. Employing a participatory plant breeding (PPB) approach may have many benefits, including increased and more stable productivity, faster release and adoption of wheat varieties, better understanding of farmers' various criteria, enhanced biodiversity, the conservation of crop diversity on farms [105,106], increased cost-effectiveness, the ability to facilitate the learning of farmers, and the empowerment of farmers [107].

Participatory plant breeding (PPB) methods incorporate the involvement of end users in the breeding process [108] and the decentralization of selection sites into farmers' fields [109]. This has been proposed as an alternative to formal plant breeding and is more likely to produce varieties acceptable to farmers in marginal environments [110]. Social studies concerning the related historical and cultural traditions can assess the needs of both farmers and the market for local products [111–113]. Usually, end users value different traits than plant breeders [114]. Plant breeders contribute their expertise in creating genetic variation, in population management, and in designing screening methods that can separate genetic from environmental effects [110]. Participation provides flexibility in the selection program. The objectives could be reoriented to ensure relevant end products in case some changes are necessary during the breeding procedure. Moreover, there is a mistrust of modern varieties among farmers, bakers, and consumers [81,115], and participatory plant breeding could rebuild client trust with improved varieties [116]. In many breeding programs where there are G x E interactions during evaluation, the lines selected under PPB have been found to perform better for farmer priority traits than those selected via formal plant breeding methods [117,118]. PPB programs can also reduce the costs of the breeding process. Cost savings primarily derive from the less frequent testing of advanced lines [119]. Many studies have shown that participatory variety selection (PVS) can improve the adoption of varieties [120–122], and thereby enhance productivity [123].

In Syria, decentralized participatory selection by farmers is significantly more efficient in identifying the highest yielding entries in farmers' fields than any other selection strategy [124]. Farmer-selected populations are not genetically homogenous, which may lead to higher yield stability in varying environments [125]. PPB projects including farmers have resulted in the wider and simpler adoption of new varieties [114,126,127]. Wheat populations after PPB will evolve by adapting gradually and continuously to climate change [128]. A wheat population developed at ICARDA [129] has been evolving for five years at a farm in Tuscany (using evolutionary participatory breeding principals); the name of this population is 'SOLIBAM Tenero Floriddia' [130]. Thanks to the EU Commission Implementing Decision (2014/150/EU), which provides specific derogations for the marketing of wheat populations, this genetically heterogeneous population is now, for the first time, being marketed as a certified seed [130].

3.3. The Application of Doubled-Haploid Techniques

The introduction of advanced in-vitro tissue culture techniques, such as androgenesis (anther or microspore culture), chromosome elimination techniques (wide hybridization), and ovule cultures (gynogenesis), in self-pollinating crop species, has helped breeders to accelerate trait fixation in segregating populations of durum wheat in research conducted in Tunisia and ICARDA [131,132]. Of the techniques mentioned above, androgenesis (more precisely, an anther culture) can only be incorporated into breeding programs if they ensure the production of a sufficient number of genetically stable doubled haploid plants from a wide range of genotypes [133]. An anther culture, despite its effectiveness and convenience, has the serious disadvantage of being firmly genotype dependent [134,135]. Furthermore, durum wheat hardly responds at all to this technique (i.e., its embryo production is deficient and most of the plants produced are albinos) [136,137]. For this reason, chromosome elimination techniques are an attractive alternative approach, since they are not genotype dependent [138] and are not influenced by the dominant Kr wheat crossover genes [139]. This technique is mainly used in producing new germplasm, not only in durum but also in bread wheat and triticale. The fourth technique, gynogenesis, is another alternative for producing new germplasm. In gynogenesis, haploid plant development is induced by an unpollinated ovary culture. However, the use of an ovary culture is practiced more rarely in wheat breeding programs [140]. In a recent study in Tunisia, Slama-Ayed et al. [132] compared three doubled haploid techniques and found that gynogenesis is an exciting approach that could be used to produce new durum wheat genotypes as a supplement to maize techniques.

4. Breeding Challenges

Durum wheat breeding is considered to be one of the most cost effective and environmentally safe ways to meet the future challenges that durum wheat productivity will face due to climate change. The durum wheat is cultivated in rain-fed farming systems in Mediterranean basin. This is mainly connected with the high temperatures and drought that are expected to become more severe the next years and affect the cultivation across the Mediterranean [141]. In this context, investment in the productivity of rain-fed areas that cover a significant portion of Mediterranean countries could contribute to food security and rural growth. The UN reports several strategies for agronomic practices based on output and productivity in semi-arid areas, including the use of adaptable varieties, which is considered a very effective practice [142]. Breeding programs must be even more efficient due to the upcoming climate change effects and increased food demands. The identification of genetic resources and the study of genetic variability will provide further information regarding the increased tolerance of durum wheat under abiotic and biotic stresses. This could contribute to the increase and stability of production in future adverse climatic conditions. In this way, genetic studies for the identification of QTLs/genes that control important agronomic traits [82,143–145] and disease resistance [146] could also help. The identification of genomic regions that affect valuable target traits is known as quantitative trait locus (QTL) (or linkage) mapping, and it is a useful tool for the exploitation of loci that are co-segregating with traits of interest in a population [147]. QTL studies have been widely conducted on durum wheat for the genetic dissection of important breeding traits using diverse molecular markers and detailed genetic maps. Maps were used for the identification of QTLs controlling several characteristics, such as grain yield and kernel characters [82,148], grain-milling traits [149], and quality traits like endosperm color [150], grain protein content [151], and other pasta quality traits [152]. In extended experiments, including 249 recombinant inbred durum wheat lines evaluated in 16 environments, it was found that two major QTLs on chromosomes 2BL and 3BS have consistent effects across different environments [82]. Also, a QTL for plant height was identified on chromosomes 1BS, 3AL, and 7AS, and three QTLs for heading date were identified on chromosomes 2AS, 2BL, and 7BS. Moreover, 76 QTLs were identified for yield components along with several morpho-physiological traits (peduncle length, the Normalized Difference Vegetation Index (NDVI), and leaf greenness at the milk-grain stage expressed in Single-photon Avalanche Diode (SPAD) units) [143]. In a study under salinity conditions, four SSR markers were closely linked with grain yield, which could thus be used in the improvement of durum wheat through MAS under abiotic stress [144].

It is also expected that climate change will affect the vulnerability of durum wheat in different diseases [153,154]. Biotic resistance has also been investigated for the identification of QTLs that confer resistance to fungi [155] or pests [156]. Additional QTL studies use even greater genetic diversity, such as multiparental crosses for the identification of yield-related QTLs [157]. Abiotic resistance has also been under investigation for the identification of QTLs. A genome-wide association study of a durum wheat core set using 7652 Single Nucleotide Polymorphism (SNP) markers allowed the identification of major QTLs controlling the adaptation to heat stress [145]. Additionally, 12 loci were found to control the main heat tolerance traits; among these loci, three activated only when heat stress occurred. Moreover, two loci validated in a Kompetitive Allele Specific PCR (KASP) marker, are ready for deployment via MAS and could result in increased productivity in heat-stressed areas and improved resilience to climate change. A haplotype analysis of 208 elite lines confirmed that those with positive allele at all three QTLs had an 8% higher yield in a heat-stressed field environment [145].

Another important parameter is the durum wheat quality in terms of its protein content, endosperm texture, and glutenin content, which cannot be easily measured phenotypically. However, the methods for testing quality are typically costly, time-consuming, and need relatively large amounts of grain, which are available only in the late stages of breeding programs. Thus, markers for wheat quality traits can be very useful to enable the screening of a high number of lines and can be used early in breeding programs [158,159]. The durum wheat breeding programs carried out over the 20th century have focused on an increase of yield in combination with quality characteristics for pasta products [52,160–162] and the achievement of better adaptability under Mediterranean conditions [52]. Little attention has been given to increasing other grain health-promoting components and nutritive constituents of durum wheat, such as dietary fibre (DF), total and soluble arabinoxylan content, and beta-glycan in semolina. There are indications that intense breeding either increases or does not affect these parameters in modern cultivars compared to old cultivars [88], indicating that the breeding process may contribute to a further improvement of durum wheat's nutritive characteristics.

Thus, markers for wheat quality traits can be very useful to enable the screening of a high number of lines and can be used early in breeding programs [158,159]. Six QTLs explained 49%–56% of grain protein variations [163], and seven QTLs explain 62%–91% of the sodium dodecyl sulfate (SDS) volume [160] in durum wheat germplasm. A number of markers targeting different glutenin alleles have been referenced, including markers for *Glu-B1* alleles, based on the sequence variations of Bx type genes [161]. Further, MAS succeeded in increasing pasta-quality-associated properties through the transfer of significant QTLs, such as the *Gli-B1* locus containing γ-gliadin 45 and the *Glu-B3* locus containing Low Molecular Weight (LMW)-2 type glutenins [162].

4.1. Seed Storage Proteins and Quality

Seed storage proteins are prolamins that account for 80% of total grain proteins, and their role is crucial in determining the technological properties of durum wheat end products [57]. Prolamins are alcohol-soluble and can be classified according to their electrophoretic mobility in two classes: monomeric gliadins and polymeric glutenins. The former can be further classified as α, β, γ, and ω gliadins or as the high and low molecular weight glutenin subunits (HMW-GS and LMW-GS respectively) [57]. Many reports have discussed the effects of gluten protein composition on durum's end products [164,165]. These effects are either genotypic or environmentally dependent [166,167]. It is well established that certain HMW and LMW glutenin subunits affect the end product quality differently in durum wheat. For example, HMW GS 7+8 alleles are associated with better quality compared to allelic HMW-GS 20 [110]. For LMW-GS, it has also been demonstrated that certain subunits encoded by the loci located on chromosome *1B* (*Glu-*) positively (LMW -2 group of subunits) or negatively (LMW-1 group of subunits) affect pasta-making properties [168].

Wheat gliadin is also characterized by high intervarietal polymorphism, and most individual cultivars show unique electrophoretic patterns [169–171]. In durum wheat, the presence of components γ-42 and γ-45 encoded by allelic genes on chromosome *1B* is reported to affect the viscoelastic properties of gluten [172]. Gliadin γ-45 is associated with a group of LMW–GS subunits termed *LMW-2*, and γ-42 is associated with LMW-1 glutenin subunits. Gliadin γ-45 could be used as a genetic marker for high gluten quality, whereas gliadin's γ-42 component could serve as a genetic marker for poor gluten quality. Also, gliadin alleles were found to be correlated with resistance to cold and stem rust [173]. Finally, in breeding programs, knowledge of the allelic composition at each locus is beneficial in identifying and using the genotypes that carry the most promising qualitative traits.

4.2. Seed Quality Characteristics Connected with Human Health

The durum wheat breeding programs carried out over the 20th century mainly focused on increasing yield in combination with quality characteristics for pasta products [52,174–176] and the achievement of better adaptability to Mediterranean conditions [52]. Little attention was given to other grain health-promoting components. It has been suggested that intensive breeding has led to decreased contents of health-promoting components in modern wheat cultivars [177]. Recently, several researchers have investigated, in detail, the other nutritive constituents of durum wheat, such as dietary fiber (DF), that have many health benefits; it was found that intense breeding has not decreased DF in the modern cultivars compared to the old ones [88]. It was also observed that the total arabinoxylan content in wholemeal or semolina is not differentiated between recent and old genotypes, while modern cultivars have higher proportions of soluble arabinoxylan in wholemeals and of beta-glycan in semolina compared to the old genotypes [88]. These results show that the breeding process could contribute to further improvements in durum wheat's nutritive characteristics.

Apart from its value as a source of nutrients, wheat may cause inflammatory immune reactions and disorders like wheat allergies, celiac disease and non-celiac wheat sensitivity (NCWS), fructose malabsorption, and irritable bowel syndrome (IBS), highlighting the need for less-reactive wheat products that can contribute to quality of life improvements [81,178]. Wheat proteins, including gluten and non-gluten proteins like amylase/trypsin inhibitors (ATI) and others, are characterized as triggering factors. Recent studies have sought to investigate the underlying causes of these immune reactions [178]. According to the types of reactions caused, wheat-related disorders are classified as: (a) allergies, including immunoglobulin E (IgE) and non-IgE mediated allergic reactions; and (b) autoimmune, including celiac disease and herpetiform dermatitis [179]. Several studies have focused on the factors that affect the immunostimulatory capacity of allergic factors present in cereals, since it has been reported that short immunotherapy may represent a valid way to treat the disease [179,180].

Recent comparative studies on the nutritional characteristics of old and modern durum wheat genotypes have found that the breeding process improves durum wheat's gluten quality both in terms of its technological performance in producing high-quality pasta products and its allergenic potential [57]. More specifically, in modern cultivars, a higher gluten index was found to be connected to increased glutenin content. Further, the breeding process contributed to the drastic reduction of a significant allergen in wheat-dependent exercise-induced anaphylaxis (WDEIA), while the old and modern durum cultivars were not different in their α-type and γ-type gliadin content, the former being considered a factor associated with celiac disease toxicity [57]. Despite the existence of allergens in wheat grain, there is genetic variability within wheat's genetic resources, and further research is necessary for the identification and the development of cultivars with lower reactivity and/or higher secondary health-promoting ingredients to meet the different needs of consumers [81,174].

5. Future Prospects

Modern genome-wide association studies (GWASs) offer the advantage of performing association analyses using the association of each marker and the phenotype of interest that has been scored across a large number of unrelated genetic materials. Furthermore, GWASs take advantage of the higher number of gene recombinations used within the panel compared to linkage mapping where meiotic recombinations are limited. As a consequence, the aim of a GWAS is to locate important QTLs for complex characteristics by employing diverse germplasm collections and modern molecular markers. GWASs are complementary to QTL mapping [181]. GWAS studies have been conducted on bread wheat to analyze important characteristics. A genome-wide association study of a durum wheat core set using 7,652 single nucleotide polymorphism (SNP) markers facilitated the identification of a major QTL controlling adaptation to heat stress [145]. Additionally, 12 loci were found to control the main heat tolerance traits; among them, three were activated only when heat stress occurred. Twenty-nine QTLs for three different yield components were identified by a GWAS in a panel of 233 tetraploid wheat accessions, including durum wheat accessions, using SNP markers [182]. GWAS is a valuable tool for breeders since broad genetic resources can be screened for market-trait associations. Germplasm collections that contain a wealth of useful genes for valuable traits such as disease resistance could be used to identify possible sources of resistance.

A GWAS that focuses on drought tolerance and 17 other agronomical traits was conducted for 493 durum wheat accessions; this study identified a putative QTL that controls drought tolerance [183]. Two QTL hotspots related to stress tolerance and yield were identified on chromosomes 2A and 2B using 6211 diversity array technology (DArTseq) SNPs on a panel consisting of 208 durum wheat lines [184]. Other GWAS studies have focused on other traits, such as disease resistance [185–188] and important quality traits. Marcotuli et al. [189] identified 37 marker-trait associations and 19 QTLs, possibly underlining arabinoxylan content in the grains of 104 tetraploid wheat genotypes. Arabinoxylans have been shown to have various health benefits. Furthermore, the co-migration of

QTLs for grain protein content and the candidate genes related to nitrogen metabolism found in a study of a durum wheat germplasm collection show that such approaches can be applied to MAS breeding schemes [190].

The rapid development of next-generation sequencing (NGS) technologies has facilitated the discovery of vast numbers of SNPs across genomes. SNP markers are now popular molecular markers because they are ubiquitous in plant genomes and are very easy and cheap to score. The high-throughput genotyping of wheat varieties is now applied routinely, especially after the construction of specific genotyping arrays and the sequencing of wheat genomes. A recently-developed genotyping array for wheat that includes 90,000 gene-associated SNPs is aiding the fast identification of genetic variation that underlines trait variation in wheat genetic materials [191]. This genotyping array contains mostly bread wheat SNPs, but also includes a large number of durum wheat SNPs. A total of 90 k genotyping wheat arrays have also helped in the construction of a detailed SNP-based genetic map based on 140 RILs developed from a cross between a wild emmer wheat population and a durum wheat cultivar [192]. More genetic maps are available today for durum wheat [193]. The durum wheat genome was only recently sequenced, revealing more valuable information about the crop's genome evolution during domestication and selection [194]. It is expected that this genome will aid in clarifying marker–trait associations and facilitate exploration of the genes underlying important characters. Durum wheat transcriptomes have also become increasingly available, thus aiding MAS breeding [195,196]. Genome sequencing is providing breeders with precise info about the nature of the genome changes in their breeding lines. Furthermore, genetic information obtained by DNA sequencing and extracted with the use of advanced bioinformatics tools will help in the application of new DNA-marker platforms and is expected to help enormously in genomics-assisted breeding for yield and quality. NGS has made possible the development of the first mutant library for wheat, which is now available publicly [197] while plenty of DNA information has been deposited in public databases accessible to scientists working on wheat all over the world.

Genotyping-by-sequencing (GBS) identifies genome polymorphism (SNP) NGS technologies, which facilitate the discovery of genetic variation in natural populations of many plants, including wheat. GBS is a useful tool and has revealed that winter durum wheat lines have significant genetic diversity, which is crucial for breeding [198]. DArTseq and SNP markers based on GBS technology were used to survey the genetic variation and the genomic characterization of 91 durum wheat landraces from Turkey and Syria, revealing extensive mixing of landraces between the two geographical regions [199]. The use of GBS in a large wheat accession collection resulted in the discovery of thousands of new SNP variations for drought and heat stress tolerance [104], which is useful for improving the elite wheat pool and enriching it with novel drought and heat tolerance genes. According to Sehgal et al. [104], this unexploited genetic variation can further improve the drought and heat stress tolerance of the elite wheat pool and enrich it with novel drought and heat tolerance genes. As the ultimate MAS tool, GBS can effectively facilitate breeding.

Furthermore, a very promising modern tool in plant breeding is genomic selection (GS). GS is a strategy used to predict the genetic value of selection candidates based on the estimated genomic breeding value, which is predicted using high-density molecular markers that are dispersed across the genome [200]. GS bases its success on the use of genome-wide markers to ensure that minor to medium effect QTLs cannot be left uncaptured unlike MAS, which focuses on the few markers linked to major genes [201]. GS models have shown high forward prediction accuracies and an enhanced genetic gain for semolina, as well as grain quality characteristics revealing that a combination of MAS and GS can be used effectively to select for quality traits [202].

The development of speed breeding is a very promising technique that could substantially help in this area. Speed breeding entails the use of specifically controlled-environment plant growth conditions and extended photoperiods of 22 h light/2 h dark that accelerates plant development. As a result, plant breeding speeds can reach up to 6–8 generations/year for wheat. Therefore, speed breeding accelerates genetic gain and significantly reduces the length of breeding cycles [203]. Specifically, studies on durum wheat [204] have recently shown that the application of constant light and controlled temperature greenhouse conditions allow the rapid growth of durum wheat seedlings and the quick phenotyping for five important traits. Early selection in the F2 generation of a bi-parental cross has led to the significant improvement of traits like crown rot tolerance, root angle, and root number, thereby proving that a combination of speed breeding with early selection can facilitate the time and efficiency of breeding programs, as recombinant inbred lines can be provided with the desirable alleles [204]. For GWAS, using speed breeding, 393 durum Recombinant Inbred Lines (RILS) and DArT-seq markers have identified a major QTL for the seminal root angle on chromosome 6A [204]. Speed breeding coupled with genomics-based technologies and other advances in phenomics could yield significant progress in the rate of genetic gain in breeding schemes.

Genetically modified wheat has been developed previously [205,206], but today no GMO wheat is cultivated officially in any part of the world. However, new technologies like genome editing and its relevant protocols provide promising tools for the future. Transcription activator-like effector nucleases (TALENs) and clustered regularly interspaced short palindromic repeat (CRISPR)–associated protein 9 (Cas9) systems have been used on bread wheat to modify three homoalleles that code for mildew-resistance locus (MLO) proteins and a TaMLO-A1 allele, respectively, to make the wheat resistant to powdery mildew [207]. Detailed improved protocols for the application of CRISPR/Cas9-mediated mutagenesis are emerging; these protocols would help achieve fast and efficient gene targeting in wheat [208–210]. Furthermore, gene editing could be used in studying gene function. Resequencing of 1526 tetraploid and 1200 hexaploid wheat mutants created a database of 10 million sequenced mutations which, by more than 90%, result in truncations or deleterious amino acid changes [211]. More precise mutations can be introduced in wheat by gene editing and this, coupled with improved transformation technologies now evolving in wheat, would mean that researchers could be further helped in their efforts to introduce novel allelic diversity for breeding durum wheat and better understand basic gene function. However, since genetic modification in Europe is legally tightly regulated and gene editing has recently been ruled out as a form of genetic modification, the production of wheat cultivars based on these techniques has to be carefully considered in the future. Nevertheless, all modern biotechnological approaches (high throughput genome analysis, gene editing, genetic engineering, and proteomics and transcriptomics) are powerful tools to complement the classical methods of breeding. It is now proposed that genome assembly, germplasm characterization, gene function identification, genomic breeding, and gene editing constitute a comprehensive 5G approach in modern breeding that could help develop new varieties with a high yield, good quality, and strong resilience to changing climate conditions [212].

6. Conclusions

Classical breeding approaches will continue to play an important role in durum wheat improvement for the release of cultivars. Advances in DNA sequencing and other technologies, such as bioinformatics, statistics, and other scientific areas, could help breeders increase the efficiency and speed of a breeding program to meet humankind's growing demands for more food that is nutritious and sustainably produced. Ultimately, the use of new molecular biology technologies is essential, but also inexorably coupled with reliable and extensive testing under real field conditions.

Author Contributions: All authors have contributed in writing this review paper. All authors have read and agreed to the published version of the manuscript

Funding: There was no specific funding for this research.

Conflicts of Interest: The authors declare that they have no conflicts of interest.

References

1. Agriculture and Agri-Food Canada. Canada: Outlook for Principal Field Crops. 19 July 2019. Available online: http://www.agr.gc.ca/eng/industry-markets-and-trade/canadian-agri-food-sector-intelligence/crops/reports-and-statistics-data-for-canadian-principal-field-crops/?id=1378743094676 (accessed on 26 September 2019).
2. EUROSTAT. Available online: https://ec.europa.eu/eurostat/data/database (accessed on 27 September 2019).
3. Agriculture and Agri-Food Canada. Area, Yield, and Production of Canadian Principal Field Crops Report. Available online: https://aimis-simia.agr.gc.ca/rp/index-eng.cfm?action=pR&r=243&lang=EN (accessed on 26 September 2019).
4. United States Department of Agriculture. Wheat Data. Available online: http://www.ers.usda.gov/data-products/wheat-data.aspx (accessed on 26 September 2019).
5. Tidiane Sall, A.; Chiari, T.; Legesse, W.; Seid-Ahmed, K.; Ortiz, R.; van Ginkel, M.; Bassi, F.M. Durum wheat (*Triticum durum* Desf.): Origin, cultivation and potential expansion in Sub-Saharan Africa. *Agronomy* **2019**, *9*, 263. [CrossRef]
6. Tedone, L.; Alhajj Ali, S.; De Mastro, G. Optimization of nitrogen in durum wheat in the Mediterranean climate: The agronomical aspect and greenhouse gas (GHG) emissions. In *Nitrogen in Agriculture—Updates*; Amanullah, K., Fahad, S., Eds.; InTech: London, UK, 2018; Volume 8, pp. 131–162.
7. Shewry, P.R. Wheat. *J. Exp. Bot.* **2009**, *60*, 1537–1553. [CrossRef]
8. Peng, J.H.; Sun, D.; Nevo, E. Domestication evolution, genetics and genomics in wheat. *Mol. Breed.* **2011**, *28*, 281–301. [CrossRef]
9. Ozkan, H.; Brandolini, A.; Schafer-Pregl, R.; Salamini, F. AFLP analysis of a collection of tetraploid wheats indicates the origin of emmer and hard wheat domestication in Southeast Turkey. *Mol. Biol. Evol.* **2002**, *19*, 1797–1801. [CrossRef] [PubMed]
10. Martínez-Moreno, F.; Solís, I.; Noguero, D.; Blanco, A.; Özberk, İ.; Nsarellah, N.; Elias, E.; Mylonas, I.; Soriano, J.M. Durum wheat in the Mediterranean Rim: Historical evolution and genetic resources. *Genet. Resour. Crop. Evol.* **2020**. [CrossRef]
11. Nalam, V.J.; Vales, M.I.; Watson, C.J.W.; Kianian, S.F.; Riera-Lizarazu, O. Map-Based analysis of genes affecting the brittle rachis character in tetraploid wheat (*Triticum turgidum* L.). *Theor. Appl. Genet.* **2006**, *112*, 373–381. [CrossRef]
12. Ben-Abu, Y.; Tzfadia, O.; Maoz, Y.; Kachanovski, D.E.; Melamed-Bessudo, C.; Feldman, M.; Levy, A.A. Durum wheat evolution: A genomic analysis. In *Série A Mediterranean Seminars, No 110, Options Méditerranéennes, Proceedings of the International Symposium on Genetics and Breeding of Durum Wheat, Rome, Italy, 27–30 May 2013*; Porceddu, E., Damania, A.B., Qualset, C.O., Eds.; CIHEAM: Bari, Italy, 2014; pp. 31–45.
13. Christidis, B. *Winter Cereals*, 2nd ed.; Thessaloniki, Greece, 1963; pp. 43–70. (In Greek)
14. Bell, G.D.H. The history of wheat cultivation. In *Wheat Breeding: Its Scientific Basis*; Lupton, F.G.H., Ed.; Springer: Dordrecht, The Netherlands, 1987; pp. 31–49.
15. Valamoti, S.M.; Gkatzogia, E.; Hristova, I.; Marinova, E. Iron age cultural interactions, plant subsistence and land use in Southeastern Europe inferred from archaeobotanical evidence of Greece and Bulgaria. In *Archaeology Across Frontiers and Borderlands: Fragmentation and Connectivity in the North Aegean and the Central Balkans from the Bronze Age to the Iron Age*; Gimatzidis, S., Pieniążek, M., Mangaloğlu-Votruba, S., Eds.; Austrian Academy of Sciences: Vienna, Austria, 2018; pp. 269–290.
16. Scarascia Mugnozza, G.T. The contribution of Italian wheat geneticists: From Nazareno Strampelli to Francesco D'Amato. In *The Wake of the Double Helix: From the Green Revolution to the Gene Revolution, Proceedings of the International Congress, Bologna, Italy, 27–31 May 2003*; Tuberosa, R., Phillips, R.L., Gale, M., Eds.; Avenue Media: Bologna, Italy, 2005; pp. 53–75.
17. Sciacca, F.; Cambrea, M.; Licciardello, S.; Pesce, A.; Sciacca, F.; Cambrea, M.; Licciardello, S.; Pesce, A.; Romano, R.; Spina, A.; et al. Evolution of durum wheat from Sicilian landraces to improved varieties. In *Série A Mediterranean Seminars, No 110, Options Méditerranéennes, Proceedings of the International Symposium on Genetics and Breeding of Durum Wheat, Rome, Italy, 27–30 May 2013*; Porceddu, E., Damania, A.B., Qualset, C.O., Eds.; CIHEAM: Bari, Italy, 2014; pp. 139–145.

18. Papadakis, I. Greek types of wheat. In *Scientific Bulletin of the "Special Station of Plant Breeding in Thessaloniki"*; General Directorate for Settlements of Makedonia: Thessaloniki, Greece, 1929; pp. 6–29. (In Greek and In French)

19. Melas, T.B. *The Finest Varieties of Greek Wheat*; Papaspyrou Printing House: Athens, Greece, 1922; pp. 33–46. (In Greek)

20. Ozberk, I.; Ozberk, F.; Atli, A.; Cetin, L.; Aydemir, T.; Keklikci, Z.; Onal, M.A.; Braun, H.J. Durum wheat in Turkey: Yesterday, today and tomorrow. In *Durum Wheat Breeding: Current Approaches and Future Strategies*; Royo, C., Nachit, M.M., Di Fonzo, N., Araus, J.L., Pfeiffer, W.H., Slafer, G.A., Eds.; The Haworth Press Inc.: Philadelphia, PA, USA, 2005; pp. 981–1010.

21. Palamarchuk, A. Selection strategies for traits relevant for winter and facultative durum wheat. In *Durum Wheat Breeding: Current Approaches and Future Strategies*; Royo, C., Nachit, M.M., Di Fonzo, N., Araus, J.L., Pfeiffer, W.H., Slafer, G.A., Eds.; The Haworth Press Inc.: Philadelphia, PA, USA, 2005; pp. 599–644.

22. Abdelkader, B. The history of wheat breeding in Algeria. In *Série A Mediterranean Seminars, No 110, Options Méditerranéennes, Proceedings of the International Symposium on Genetics and Breeding of Durum Wheat, Rome, Italy, 27–30 May 2013*; Porceddu, E., Damania, A.B., Qualset, C.O., Eds.; CIHEAM: Bari, Italy, 2014; pp. 363–370.

23. Taghouti, M.; Nsarellah, N.; Rhrib, K.; Benbrahim, N.; Amallah, L.; Rochdi, A. Evolution from durum wheat landraces to recent improved varieties in Morocco in terms of productivity increase to the detriment of grain quality. *Rev. Mar. Sci. Agron. Vét.* **2017**, *5*, 351–358.

24. Boukida, F.; Vittadinib, E.; Prandib, B.; Mattarozzid, M.; Marchinib, M.; Sforzab, S.; Sayare, R.; WeonSeof, Y.; Yacoubia, I.; Mejri, M. Insights into a century of breeding of durum wheat in Tunisia: The properties of flours and starches isolated from landraces, old and modern genotypes. *Food Sci. Technol.* **2018**, *97*, 743–751. [CrossRef]

25. Kadkol, G.P.; Sissons, M. Durum wheat: Overview. In *Encyclopedia of Food Grains*, 2nd ed.; Wrigley, C., Corke, H., Seetharaman, K., Faubion, J., Eds.; Academic Press: Oxford, UK, 2016; pp. 117–124.

26. Clarke, J.M.; De Ambrogio, E.; Hare, R.A.; Roumet, P. Genetics and breeding of durum wheat. In *Durum Wheat Chemistry and Technology*, 2nd ed.; Sissons, M., Abecassis, J., Marchylo, B., Carcea, M., Eds.; AACC International: St. Paul, MN, USA, 2012; pp. 15–36.

27. Porceddu, E.; Blanco, A. Evolution of durum wheat breeding in Italy. In *Série A Mediterranean Seminars, No 110, Options Méditerranéennes, Proceedings of the International Symposium on Genetics and Breeding of Durum Wheat, Rome, Italy, 27–30May 2013*; Porceddu, E., Damania, A.B., Qualset, C.O., Eds.; CIHEAM: Bari, Italy, 2014; pp. 157–173.

28. Bagnara, D.; Scarascia Mugnozza, G.T. Outlook in breeding for yield in durum wheat. In *Genetics and Breeding of Durum Wheat, Proceedings of a Symposium on Genetics and Breeding of Durum Wheat, Bari, Italy, 14–18 May 1973*; Scarascia Mugnozza, G.T., Ed.; Eucarpia: Wageningen, The Netherlands, 1975; pp. 249–273.

29. Papadakis, I. *The Plant: Breeding Institute 1923–1933*; Bulletin No. 15; Kastrinaki & Georganta: Thessaloniki, Greece, 1933; pp. 1–36. (In Greek)

30. Percival, J. *The Wheat Plant*; A Monograph; Duckworth: London, UK, 1921; pp. 1–628, A Monograph.

31. Papadakis, I. *List of Seeds: What Seeds Should We Plant*; Central Committee for the Protection of Domestic Seed Production, Plant Breeding Institute: Thessaloniki, Greece, 1934; pp. 1–10. (In Greek)

32. Kokolios, B. Durum wheat in Greece. In *Proceedings of Convegno Internationale del Grano Duro*; Camera di Commercio Industria de Agricoltura: Foggia, Italy, 1962.

33. Di Fonzo, N.; Blanco, A.; Ravaglia, S.; Troccoli, A.; DeAmbrogio, E. Durum wheat improvement in Italy. In *Durum Wheat Breeding: Current Approaches and Future Strategies*; Royo, C., Nachit, M.N., Di Fonzo, N., Araus, J.L., Pfeiffer, W.H., Slafer, G.A., Eds.; The Haworth Press Inc.: Philadelphia, PA, USA, 2005; pp. 825–881.

34. Royo, C.; Briceño-Félix, G.A. Wheat breeding in Spain. In *The World Wheat Book: A History of Wheat Breeding*; Angus, W., Bonjean, A., Van Ginkel, M., Eds.; Lavoisier Publishing Inc.: Paris, France, 2011; Volume 2, pp. 121–154.

35. Royo, C. Durum wheat improvement in Spain. In *Durum Wheat Breeding: Current Approaches and Future Strategies*; Royo, C., Nachit, M.N., Di Fonzo, N., Araus, J.L., Pfeiffer, W.H., Slafer, G.A., Eds.; The Haworth Press Inc.: Philadelphia, PA, USA, 2005; pp. 883–906.

36. Grignac, P. Action of leaf rust (Puccinia recondita f. sp. Tritici) on durum wheat yield and breeding for rust resistance. In *Genetics and Breeding of Durum Wheat, Proceedings of a Symposium on Genetics and Breeding of Durum Wheat, Bari, Italy, 14–18 May 1973*; Scarascia Mugnozza, G.T., Ed.; Eucarpia: Wageningen, The Netherlands, 1975; pp. 489–496.

37. Braun, H.J.; Zencirci, N.; Altay, F.; Atli, A.; Avci, M.; Eser, V.; Kambertay, M.; Payne, T.S. Turkish wheat pool. In *The World Wheat Book: A History of Wheat Breeding*; Bonjean, A.P., Angus, W.J., Eds.; Lavoisier: Paris, France, 2001; pp. 851–879.

38. Taghouti, M.; Nsarellah, N.; Gaboun, F.; Rochdi, A. Multi-Environment assessment of the impact of genetic improvement on agronomic performance and on grain quality traits in Moroccan durum wheat varieties of 1949 to 2017. *Glob. J. Plant Breed. Genet.* **2017**, *4*, 394–404.

39. Byerlee, D.; Moya, P. *Impacts of International Wheat Breeding Research in the Developing World, 1966–1990*; The International Maize and Wheat Improvement Center (CIMMYT): Mexico City, Mexico, 1993; pp. 69–70.

40. Salazar, J.; Branas, M.; Ziteli, G.; Vallega, J. Old Iberian durum wheats as an important source of resistance to stem and leaf rusts. In *Genetics and Breeding of Durum Wheat, Proceedings of a Symposium on Genetics and Breeding of Durum Wheat, Bari, Italy, 14–18May 1973*; Scarascia Mugnozza, G.T., Ed.; Eucarpia: Wageningen, The Netherlands, 1975; pp. 497–508.

41. Al-Idrissi, M.; Sbeita, A.; Jebriel, S.; Zintani, A.; Shreidi, A.; Ghawawi, H.; Tazi, M. *Libya: Country Report to the FAO International Technical Conference on Plant Genetic Resources*; The Food and Agriculture Organization of the United Nations(FAO): Leipzig, Germany, 1996.

42. Ghanem, E.H. Wheat improvement in Egypt with emphasis on heat tolerance. In *Wheat in Heat-Stressed Environments: Irrigated, Dry Areas and Rice-Wheat Farming Systems*; Saunders, D.A., Hettel, G.P., Eds.; The International Maize and Wheat Improvement Center(CIMMYT): Mexico City, Mexico, 1994.

43. Zencirci, N.; Kinaci, E.; Atli, A.; Kalayci, M.; Avci, M. Wheat research in Turkey. In *Wheat: Prospects for Global Improvement. Developments in Plant Breeding*; Braun, H.J., Altay, F., Kronstad, W.E., Beniwal, S.P.S., McNab, A., Eds.; Springer: Dordrech, The Netherlands, 1997; Volume 6, pp. 11–16.

44. Poiarkova, H.; Blum, A. Land-Races of wheat from the northern Negev in Israel. *Euphytica* **1983**, *32*, 257–271. [CrossRef]

45. Isidro, J.; Álvaro, F.; Royo, C.; Villegas, D.; Miralles, D.J.; García del Moral, L.F. Changes in Duration of Developmental Phases of Durum Wheat Caused by Breeding in Spain and Italy during the 20th Century and Its Impact on Yield. *Ann. Bot.* **2011**, *107*, 1355–1366. [CrossRef] [PubMed]

46. Hedden, P. The genes of the Green Revolution. *Trends Genet.* **2003**, *19*, 5–9. [CrossRef]

47. Lebsock, K.L. Transfer of Norin 10 genes for dwarfness to durum wheat. *Crop Sci.* **1963**, *3*, 450–451. [CrossRef]

48. Pfeiffer, W.H.; Sayre, K.D.; Reynolds, M.P. Enhancing genetic grain yield potential and yield stability in durum wheat. In *Durum Wheat Improvement in the Mediterranean Region: New Challenges*; Royo, C., Nachit, M., Di Fonzo, N., Araus, J.L., Eds.; International Centre for Advanced Mediterranean Agronomic Studies(CIHEAM): Zaragoza, Spain, 2000; pp. 83–93.

49. The International Center for Agriculture Research in the Dry Areas(ICARDA). Available online: https://www.icarda.org/ (accessed on 30 August 2019).

50. Bassi, F.M.; Nachit, M.M. Genetic gain for yield and allelic diversity over 35 years of durum wheat breeding at ICARDA. *Crop Breed. Genet. Genom.* **2019**, *1*, 1–19.

51. McCaig, T.N.; Clarke, J.M. Breeding durum wheat in western Canada: Historical trends in yield and related variables. *Can. J. Plant Sci.* **1995**, *75*, 55–60. [CrossRef]

52. De Vita, P.; Matteu, L.; Mastrangelo, A.M.; Di Fonzo, N.; Cattivelli, L. Effects of breeding activity on durum wheat traits breed in Italy during the 20th century. *Ital. J. Agron.* **2007**, 451–462. [CrossRef]

53. Waddington, S.R.; Osmanzai, M.; Yoshida, M.; Ransom, J.K. The Yield of durum wheats released in Mexico between 1960 and 1984. *J. Agric. Sci.* **1987**, *108*, 469–477. [CrossRef]

54. Royo, C.; Álvaro, F.; Martos, V.; Ramdani, A.; Isidro, J.; Villegas, D.; García del Moral, L.F. Genetic changes in durum wheat yield components and associated traits in Italian and Spanish varieties during the 20th century. *Euphytica* **2007**, *155*, 259–270. [CrossRef]

55. Motzo, R.; Fois, S.; Giunta, F. Relationship between grain yield and quality of durum wheats from different eras of breeding. *Euphytica* **2004**, *140*, 147–154. [CrossRef]

56. Guarda, G.; Padovan, S.; Delogu, G. Grain yield, nitrogen-use efficiency and baking quality of old and modern Italian bread-wheat cultivars grown at different nitrogen levels. *Eur. J. Agron.* **2004**, *21*, 181–192. [CrossRef]

57. De Santis, M.A.; Giuliani, M.M.; Giuzio, L.; De Vita, P.; Lovegrove, A.; Shewry, P.R.; Flagella, Z. Differences in gluten protein composition between old and modern durum wheat genotypes in relation to 20th century breeding in Italy. *Eur. J. Agron.* **2017**, *87*, 19–29. [CrossRef] [PubMed]

58. Institute of Plant Breeding and Genetic Resources. Available online: http://http://www.ipgrb.gr/index.php/antikeimena/sitiron (accessed on 27 October 2019).

59. Kotzamanidis, S. Three new cultivars of durum wheat with very high yield potential, quality and excellent adaptation to Greek conditions. *ETHIAGE* **2011**, *43*, 14–15. (In Greek)

60. Korpetis, E.; Bladenopoulos, K.; Kotzamanidis, S. Two new Greek cultivars of Cereal Institute: THRAKI (durum wheat) and ALKMINI (triticale). In Proceedings of the 15th National Conference of Hellenic Scientific Society for Genetics and Plant Breeding, Larisa, Greece, 15–17 October 2014; p. 157. (In Greek)

61. Chairi, F.; Vergara-Diaz, O.; Vatter, T.; Aparicio, N.; Nieto-Taladriz, M.T.; Kefauver, S.C.; Bort, J.; Serret, M.D.; Araus, J.L. Post-Green revolution genetic advance in durum wheat: The case of Spain. *Field Crop. Res.* **2018**, *228*, 158–169. [CrossRef]

62. Nsarellah, N.; Amri, A. Durum wheat genetic improvement in Morocco. In *Durum Wheat Breeding: Current Approaches and Future Strategies*; Royo, C., Nachit, M.N., Di Fonzo, N., Araus, J.L., Pfeiffer, W.H., Slafer, G.A., Eds.; The Haworth Press Inc.: Philadelphia, PA, USA, 2005; pp. 963–980.

63. Scarascia-Mugnozza, G.T.; D'Amato, F.; Avanzi, S.; Bagnara, D.; Belli, M.L.; Bozzini, A.; Brunori, A.; Cervigni, T.; Devreux, M.; Donini, B.; et al. Mutation breeding for durum wheat (*Triticum turgidum* ssp. *durum* Desf.) improvement in Italy. *Mut. Breed. Rev.* **1993**, *10*, 1–28.

64. Skorda, E. The Mutability of Some Greek Varieties of the Genus Triticum. Ph.D. Thesis, Agricultural University of Athens, Athens, Greece, 1971.

65. Klatt, A.R.; Dincer, N.; Yakar, K. Problems associated with breeding spring and winter durums in Turkey. In *Genetics and Breeding of Durum Wheat, Proceedings of a Symposium on Genetics and Breeding of Durum Wheat, Bari, Italy, 14–18May 1973*; Scarascia Mugnozza, G.T., Ed.; Eucarpia: Wageningen, The Netherlands, 1975; pp. 327–337.

66. Yaniv, E.; Raats, D.; Ronin, Y.; Korol, A.B.; Grama, A.; Bariana, H.; Dubcovsky, J.; Schulman, A.H. Evaluation of marker-assisted selection for the stripe rust resistance gene Yr15, introgressed from wild emmer wheat. *Mol. Breed.* **2015**, *35*. [CrossRef]

67. Phan, H.T.T.; Rybak, K.; Bertazzoni, S.; Furuki, E.; Dinglasan, E.; Hickey, L.T.; Oliver, R.P.; Tan, K.C. Novel sources of resistance to septoria nodorum blotch in the vavilov wheat collection identified by genome-wide association studies. *Theor. Appl. Genet.* **2018**, *131*, 1223–1238. [CrossRef]

68. Vishwakarma, M.K.; Mishra, V.K.; Gupta, P.K.; Yadav, P.S.; Kumar, H.; Joshi, A.K. Introgression of the high grain protein gene Gpc-B1 in an elite wheat variety of indo-gangetic plains through marker assisted backcross breeding. *Curr. Plant Biol.* **2014**, *1*, 60–67. [CrossRef]

69. Yildirim, A.; Sonmezoglu, O.A.; Sayaslan, A.; Koyuncu, M.; Gulec, T.E.; Kanderim, N. Marker-Assisted breeding of a durum wheat cultivar for γ-gliadin and LMW-glutenin proteins affecting pasta quality. *Turk. J. Agric. For.* **2013**, *37*, 527–533. [CrossRef]

70. Pignone, D.; De Paola, D.; Rapanà, N.; Janni, M. Single seed descent: A tool to exploit durum wheat (*Triticum durum* Desf.) genetic resources. *Genet. Resour. Crop Evol.* **2015**, *62*, 1029–1035. [CrossRef]

71. Farag, H.I.A. Efficiency of three methods of selection in wheat breeding under saline stress conditions. *Egypt. J. Plant Breed.* **2013**, *17*, 85–95. [CrossRef]

72. Aoun, M.; Kolmer, J.A.; Rouse, M.N.; Elias, E.M.; Breiland, M.; Bulbula, W.D.; Chao, S.; Acevedo, M. Mapping of novel leaf rust and stem rust resistance genes in the Portuguese durum wheat landrace. *G3 Genes Genomes Genet.* **2019**, *9*, 2535–2547. [CrossRef] [PubMed]

73. Atkins, R.E.; Murphy, H.C. Evaluation of yield potentialities of oat crosses from bulk hybrid tests. *Agron. J.* **1949**, *41*. [CrossRef]

74. Fowler, W.L.; Heyne, E.G. Evaluation of bulk hybrid tests for predicting performance of pure line selections in hard red winter wheat. *Agron. J.* **1955**, *47*, 430–434. [CrossRef]

75. Lupton, F.G.H. Studies in the breeding of self-pollinating cereals. *Euphytica* **1961**, *10*, 209–224.

76. Schut, J.W.; Dourleijn, C.J.; Bos, I. Cross and Line Prediction in Barley Using F_4 Small-Plot Yield Trials. Ph.D. Thesis, Wageningen Agricultural University (WAU), Wageningen, The Netherlands, 1998; pp. 21–43.

77. Mitchell, J.W.; Baker, R.J.; Knott, D.R. Evaluation of honeycomb selection for single plant yield in durum wheat. *Crop Sci.* **1982**, *22*, 840–843. [CrossRef]

78. Kotzamanidis, S.T.; Lithourgidis, A.S.; Mavromatis, A.G.; Chasioti, D.I.; Roupakias, D.G. Prediction criteria of promising F3 populations in durum wheat: A comparative study. *Field Crop. Res.* **2008**, *107*, 257–264. [CrossRef]

79. Lungu, D.M.; Kaltsikes, P.J.; Larter, E.N. Honeycomb selection for yield in early generations of spring wheat. *Euphytica* **1987**, *36*, 831–839. [CrossRef]

80. Mylonas, I. Experimentation of the Maximization of Yield Potential in Barley (*Hordeum vulgare* L.) Cultivars. Ph.D. Thesis, School of Agriculture, Aristotle University of Thessaloniki, Thessaloniki, Greece, 2012.

81. Kissing Kucek, L.; Veenstra, L.D.; Amnuaycheewa, P.; Sorrells, M.E. A Grounded Guide to Gluten: How Modern Genotypes and Processing Impact Wheat Sensitivity. *Compr. Rev. Food Sci. Food Saf.* **2015**, *14*, 285–302. [CrossRef]

82. Maccaferri, M.; Sanguineti, M.C.; Corneti, S. Quantitative trait loci for grain yield and adaptation of durum wheat (*Triticum durum* Desf.) across a wide range of water availability. *Genetics* **2008**, *178*, 489–511. [CrossRef]

83. De Vita, P.; Nicosia, O.L.D.; Nigro, F.; Platani, C.; Riefolo, C.; Di Fonzo, N.; Cattivelli, L. Breeding progress in morpho-physiological, agronomical and qualitative traits of durum wheat cultivars released in Italy during the 20th century. *Eur. J. Agron.* **2007**, *26*, 39–53. [CrossRef]

84. Garcia del Moral, L.F.; Royo, C.; Slafer, G.A. Genetic improvement effects on durum wheat yield physiology. In *Durum Wheat Breeding: Current Approaches and Future Strategies*; Royo, C., Nachit, M.N., Di Fonzo, N., Araus, J.L., Pfeiffer, W.H., Slafer, G.A., Eds.; The Haworth Press Inc.: Philadelphia, PA, USA, 2005; pp. 379–396.

85. Araus, J.L.; Slafer, G.A.; Reynolds, M.P.; Royo, C. Plant breeding and drought in C3 cereals: What to breed for? *Ann. Bot.* **2002**, *89*, 925–940. [CrossRef] [PubMed]

86. Richards, R.A. Selectable traits to increase crop photosynthesis and yield of grain crops. *J. Exp. Bot.* **2000**, *51*, 447–458. [CrossRef] [PubMed]

87. Spielmeyer, W.; Hyles, J.; Joaquim, P.; Azanza, F.; Bonnett, D.; Ellis, M.E.; Moore, C.; Richards, R.A. A QTL on chromosome 6A in bread wheat (*Triticum aestivum*) is associated with longer coleoptiles, greater seedling vigour and final plant height. *Theor. Appl. Genet.* **2007**, *115*, 59–66. [CrossRef] [PubMed]

88. De Santis, M.A.; Kozik, O.; Passmore, D.; Flagella, Z.; Shwewry, P.R. Comparison of the dietary fibre composition of old and modern durum wheat (*Triticum turgidum* spp. *durum*) genotypes. *Food Chem.* **2018**, *244*, 304–310. [CrossRef]

89. Achtar, S.; Moualla, M.Y.; Kalhout, A.; Röder, M.S.; MirAli, N. Assessment of genetic diversity among syrian durum (*Triticum* ssp. *durum*) and bread wheat (*Triticum aestivum* L.) using SSR markers. *Russ. J. Genet.* **2010**, *46*, 1320–1326. [CrossRef]

90. Casassola, A.; Brammer, S.P.; Chaves, M.S.; Wiethölter, P.; Caierão, E. Parental selection of wheat lines based on phenotypic characterization and genetic diversity. *Crop Breed. Appl. Biotechnol.* **2013**, *13*, 49–58. [CrossRef]

91. Marzario, S.; Logozzo, G.; David, J.L.; Zeuli, P.S.; Gioia, T. Molecular genotyping (SSR) and agronomic phenotyping for utilization of durum wheat (*Triticum Durum* Desf.) ex situ collection from Southern Italy: A combined approach including pedigreed varieties. *Genes* **2018**, *9*, 465. [CrossRef]

92. Soriano, J.M.; Villegas, D.; Sorrells, M.E.; Royo, C. Durum wheat landraces from east and west regions of the mediterranean basin are genetically distinct for yield components and phenology. *Front. Plant Sci.* **2018**, *9*, 80. [CrossRef]

93. Abu-Zaitoun, S.Y.; Chandrasekhar, K.; Assili, S.; Shtaya, M.J.; Jamous, R.M.; Mallah, O.B.; Nashef, K.; Sela, H.; Distelfeld, A.; Alhajaj, N.; et al. Unlocking the genetic diversity within a Middle-East panel of durum wheat landraces for adaptation to semi-arid climate. *Agronomy* **2018**, *8*, 233. [CrossRef]

94. Dograr, N.; Akin-Yalin, S.; Akkaya, M. Discriminating durum wheat cultivars using highly polymorphic simple sequence repeat DNA markers. *Plant Breed.* **2000**, *119*, 360–362. [CrossRef]

95. Eujayl, I.; Sorrells, M.; Baum, M.; Wolters, P.; Powell, W. Isolation of EST-derived microsatellite markers for genotyping the A and B genomes of wheat. *Theor. Appl. Genet.* **2002**, *104*, 399–407. [CrossRef] [PubMed]

96. Incirli, A.; Akkaya, M.S. Assessment of genetic relationships in durum wheat cultivars using AFLP markers. *Genet. Resour. Crop Evol.* **2001**, *48*, 233–238. [CrossRef]

97. Pujar, S.; Tamhankar, S.; Rao, V.; Gupta, V.; Naik, S.; Ranjekar, P. Arbitrarily primed-PCR based diversity assessment reflects hierarchical groupings of Indian tetraploid wheat genotypes. *Theor. Appl. Genet.* **1999**, *99*, 868–876. [CrossRef]

98. Soleimani, V.; Baum, B.; Johnson, D. AFLP and pedigree-based genetic diversity estimates in modern cultivars of durum wheat (*Triticum turgidum* L. subsp. *durum* (Desf.) Husn.). *Theor. Appl. Genet.* **2002**, *104*, 350–357. [CrossRef] [PubMed]

99. Babay, E.; Mnasri, S.R.; Mzid, R.; Naceur, M.B.; Hanana, M. Quality selection and genetic diversity of tunisian durum wheat varieties using SSR markers. *Biosci. J.* **2019**, *35*. [CrossRef]

100. Bonnett, D.G.; Rebetzke, G.J.; Spielmeyer, W. Strategies for efficient implementation of molecular markers in wheat breeding. *Mol. Breed.* **2005**, *15*, 75–85. [CrossRef]

101. Kuchel, H.; Fox, R.; Hollamby, G.; Reinheimer, J.L.; Jefferies, S.P. The challenges of integrating new technologies into a wheat breeding programme. In Proceedings of the 11th International Wheat Genet Symposium, Brisbane, Australia, 24–29 August 2008; Appels, R., Eastwood, R., Lagudah, E., Langridge, P., Mackay, M., McIntyre, L., Sharp, P., Eds.; Sydney University Press: Sydney, Australia, 2008; pp. 1–5.

102. Jia, Z.; Liu, Y.; Gruber, B.D.; Neumann, K.; Kilian, B.; Graner, A.; von Wirén, N. Genetic dissection of root system architectural traits in spring barley. *Front. Plant Sci.* **2019**, *10*, 400. [CrossRef]

103. Bhullar, N.K.; Zhang, Z.; Wicker, T.; Keller, B. Wheat gene bank accessions as a source of new alleles of the powdery mildew resistance gene Pm3: A large scale allele mining project. *BMC Plant Biol.* **2010**, *10*, 88. [CrossRef]

104. Sehgal, D.; Vikram, P.; Sansaloni, C.P.; Ortiz, C.; Pierre, C.S.; Payne, T.; Ellis, M.; Amri, A.; Petroli, C.D.; Wenzl, P.; et al. Exploring and mobilizing the gene bank biodiversity for wheat improvement. *PLoS ONE* **2015**, *10*, e0132112. [CrossRef]

105. Cleveland, D.A.; Soleri, D.; Smith, S.E. Do folk crop varieties have a role in sustainable agriculture? Incorporating folk varieties into the development of locally based agriculture may be the best approach. *Bioscience* **1994**, *44*, 740–751. [CrossRef]

106. Love, B.; Spaner, D. Agrobiodiversity: Its value, measurement, and conservation in the context of sustainable agriculture. *J. Sustain. Agric.* **2007**, *31*, 53–82. [CrossRef]

107. Sperling, L.; Ashby, J.A.; Smith, M.E.; Weltzien, E.R.; McGuire, S. A framework for analyzing participatory plant breeding approaches and results. *Euphytica* **2001**, *136*, 21–35.

108. Witcombe, J.R.; Joshi, K.D.; Gyawali, S.; Musa, A.M.; Johansen, C.; Virk, D.S.; Sthapit, B.R. Participatory plant breeding is better described as highly client-oriented plant breeding. I. Four indicators of client-orientation in plant breeding. *Exp. Agric.* **2005**, *41*, 299–319. [CrossRef]

109. Ceccarelli, S. Efficiency of plant breeding. *Crop Sci.* **2015**, *55*, 87–97. [CrossRef]

110. Atlin, G.N.; Cooper, M.; Bjornstad, A. A comparison of formal and participatory breeding approaches using selection theory. *Euphytica* **2001**, *122*, 463–475. [CrossRef]

111. Pretty, J.; Vodouhê, S.D. Using rapid or participatory rural appraisal. In *Improving Agricultural Extension*; A Reference Manual; The Food and Agriculture Organization of the United Nations (FAO): Rome, Italy, 1997.

112. Soleri, D.; Cleveland, D. Breeding for quantitative variables, Part 1: Farmers' and scientists' knowledge and practice in variety choice and plant selection. In *Plant Breeding and Farmer Participation*; Ceccarelli, S., Guimarães, E.P., Weltzien, E., Eds.; The Food and Agriculture Organization of the United Nations (FAO): Rome, Italy, 2009; pp. 324–366.

113. Organic Seed Alliance (OSA). *Participatory Plant Breeding Toolkit*; Organic Seed Alliance (OSA): Port Townsend, WA, USA, 2012.

114. Ashby, J.A. The impact of participatory plant breeding. In *Plant Breeding and Farmer Participation*; Ceccarelli, S., Guimarães, E.P., Weltzien, E., Eds.; The Food and Agriculture Organization of the United Nations (FAO): Rome, Italy, 2009; pp. 649–671.

115. Davis, W. *Wheat Belly: Lose the Wheat, Lose the Weight, and Find: Your Path Back to Health*; Emmaus, P.A., Ed.; Rodale Press: Emmaus, PA, USA, 2011.

116. Kissing Kucek, L.; Darby, H.; Mallory, E.; Dawson, J.; Davis, M.; Dyck, E.; Lazor, J.; O'Donnell, S.; Mudge, S.; Kimball, M.; et al. Participatory breeding of wheat for organic production. In Proceedings of the Organic Agriculture Research Symposium, LaCrosse, WI, USA, 25–26 February 2015; pp. 25–26.

117. Goldringer, I. Co-Constructiuon of a French program for participatory breeding of wheat cultivars adapted to organic cultivation and transformation. In Proceedings of the EGOSGN Conference, Montreal, QC, Canada, 4–6 July 2014.

118. Kirk, A.P.; Fox, S.L.; Entz, M.H. Comparison of organic and conventional selection environments for spring wheat. *Plant Breed.* **2012**, *131*, 687–694. [CrossRef]

119. Mangione, D.; Senni, S.; Puccioni, M.; Grando, S.; Ceccarelli, S. The cost of participatory barley breeding. *Euphytica* **2006**, *150*, 289–306. [CrossRef]

120. Ceccarelli, S. Specific adaptation and breeding for marginal conditions. *Euphytica* **1994**, *77*, 205–219. [CrossRef]

121. Ceccarelli, S. Adaptation to low/high input cultivation. *Euphytica* **1996**, *92*, 203–214. [CrossRef]

122. Ceccarelli, S.; Erskine, W.; Hamblin, J.; Grando, S. Genotype by environment interaction and international breeding programmes. *Exp. Agric.* **1994**, *30*, 177–187. [CrossRef]

123. Ceccarelli, S.; Grando, S.; Baum, M. Participatory plant breeding in water-limited environments. *Exp. Agric.* **2007**, *43*, 411–435. [CrossRef]

124. Ceccarelli, S.; Grando, S.; Bailey, E.; Amri, A.; El-Felah, M.; Nassif, F. Farmer participation in barley breeding in Syria, Morocco and Tunisia. *Euphytica* **2001**, *122*, 521–536. [CrossRef]

125. Entz, M.H.; Kirk, A.P.; Carkner, M.; Vaisman, I.; Fox, S.L. Evaluation of lines from a farmer participatory organic wheat breeding program. *Crop Sci.* **2018**, *58*, 2433. [CrossRef]

126. Ortiz-Perez, R.; Rios-Labrada, H.; Miranda-Lorigados, S.; Ponce-Brito, M.; Quintero-Fernandez, E.; Chaveco-Perez, O. Avances del mejoramiento genetico participativo del frijol en Cuba. *Agron. Mesoam.* **2006**, *17*, 337–346. [CrossRef]

127. Mustafa, Y.; Grando, S.; Ceccarelli, S. *Assessing the Benefits and Costs of Participatory and Conventional Barley Breeding Programs in Syria*; The International Development Research: Ottawa, ON, Canada, 2006.

128. Ceccarelli, S. Drought. In *Plant Genetic Resources and Climate Change*; Jackson, M., Ford-Lloyd, B.V., Parry, M.L., Eds.; CAB International: Boston, MA, USA, 2014; pp. 221–235.

129. Ceccarelli, S. Landraces: Importance and use in breeding and environmentally friendly agronomic systems. In *Agrobiodiversity Conservation: Securing the Diversity of Crop Wild Relatives and Landraces*; Maxted, N., Dulloo, M.E., Ford-Lloyd, B.V., Frese, L., Iriondo, J., Pinheiro de Carvalho, M.A.A., Eds.; CABI Publishing: New York, NY, USA, 2012; pp. 103–117.

130. Petitti, M.; Bocci, R.; Bussi, B.; Ceccarelli, S.; Spillane, C.; McKeown, P. Future proofing decentralised evolutionary wheat populations' seed systems in Italy using a climate analogues approach: The example of Tuscany. In Proceedings of the EUCARPIA Symposium on Breeding for Diversification, Witzenhausen, Germany, 19–21 February 2018.

131. Tasesse, W.; Sanchez-Garcia, M.; Tawkaz, S.; Baum, M. Doubled haploid production in wheat. In *Advances in Breeding Techinques for Cereal Crops*; Ordon, F., Friedt, W., Eds.; Burleigh Dodds Science Publishing: Cambridge, UK, 2019; pp. 93–116.

132. Slama-Ayed, O.; Bouhaouel, I.; Ayed, S.; Jacques De Buyser, J.; Picard, E.; Hajer Slim Amara, H.S. Efficiency of three haplomethods in durum wheat (*Triticum turgidum* subsp. *durum* Desf.): Isolated microspore culture, gynogenesis and wheat × maize crosses. *Czech J. Genet. Plant Breed.* **2019**, *55*. [CrossRef]

133. Snape, J.; Simpson, E. The genetical expectations of doubled haploid lines derived from different filial generations. *Theor. Appl. Genet.* **1981**, *60*, 123–128. [CrossRef]

134. Almouslem, A.B.; Jauhar, P.P.; Peterson, T.S.; Bommineni, V.R.; Rao, M.B. Haploid durum wheat production via hybridization with maize. *Crop Sci.* **1998**, *38*, 1080–1087. [CrossRef]

135. Niroula, R.K.; Bimb, H.P. Overview of wheat x maize system of crosses for dihaploid induction in wheat. *World Appl. Sci. J.* **2009**, *7*, 1037–1045.

136. Foroughi-Wehr, B.; Zeller, F.J. In-Vitro microspore reaction of different German wheat cultivars. *Theor. Appl. Genet.* **1990**, *79*, 77–80. [CrossRef]

137. Loschenberger, F.; Hansel, H.; Heberle-Bors, E. Plant formation from pollen of tetraploid wheat species and Triticum tauschii. *Cereal Res. Commun.* **1993**, *21*, 133–139.

138. Hussain, M.; Niaz, M.; Iqbal, M.; Iftikhar, T.; Ahmad, J. Emasculation techniques and detached tiller culture in wheat x maize crosses. *J. Agric. Res.* **2012**, *50*, 1–19.

139. Laurie, D.A. Factors affecting fertilization frequency in crosses of *Triticum aestivum* cv. "Highbury" x *Zea mays* cv. "Seneca 60". *PlantBreed.* **1989**, *103*, 133–140. [CrossRef]

140. Sibi, M.L.; Kobaissi, A.; Shekafandeh, A. Green haploid plants from unpollinated ovary culture in tetraploid wheat (*Triticum durum* Defs.). *Euphytica* **2001**, *122*, 351–359. [CrossRef]

141. Gibbon, D. Rainfed farming systems in the Mediterranean region. *Plant Soil* **1981**, *58*, 59–80. [CrossRef]

142. United Nations (UN). *Global Drylands: A UN System-Wide Response*; United Nations (UN): New York, NY, USA, 2011.

143. Graziani, M.; Maccaferri, M.; Royo, C.; Salvatorelli, F.; Tuberosa, R. QTL dissection of yield components and morpho-physiological traits in a durum wheat elite population tested in contrasting thermo-pluviometric conditions. *Crop Pasture Sci.* **2014**, *65*, 80–95. [CrossRef]

144. Dura, S.; Duwayri, M.; Nachit, M.; Sheyaboudi, F.A. Detection of molecular markers associated with yield and yield components in durum wheat (*Triticum turgidum* L. Var. *durum*Desf.) under drought conditions. *Crop Pasture Sci.* **2013**, *64*, 957–964. [CrossRef]

145. Hassouni, K.E.; Belkadi, B.; Sall, A.T.; Filali-Maltouf, A.; Al-Abdallat, A.; Nachit, M.M.; Bassi, F.M. Loci controlling adaptation to heat stress occurring at the reproductive stage in durum wheat. *Agronomy* **2019**, *9*, 414. [CrossRef]

146. Kthiri, D.; Loladze, A.; MacLachlan, P.R.; N'Diaye, A.; Walkowiak, S.; Nilsen, K.; Dreisigacker, S.; Ammar, K.; Pozniak, C.J. Characterization and mapping of leaf rust resistance in four durum wheat cultivars. *PLoS ONE* **2018**, *13*, e0197317. [CrossRef]

147. Alqudah, A.M.; Sallam, A.; Baenziger, P.S.; Börner, A. GWAS: Fast-Forwarding gene identification and characterization in temperate Cereals: Lessons from Barley—A review. *J. Adv. Res.* **2020**, *22*, 119–135. [CrossRef]

148. Patil, R.M.; Tamhankar, S.A.; Oak, M.D.; Raut, A.L.; Honrao, B.K.; Rao, V.S.; Misra, S.C. Mapping of QTL for agronomic traits and kernel characters in durum wheat (*Triticum durum* Desf.). *Euphytica* **2013**, *190*, 117–129. [CrossRef]

149. Elouafi, I.; Nachit, M.M. A genetic linkage map of the durum×triticum dicoccoides backcross population based on SSRs and AFLP markers, and QTL analysis for milling traits. *Theor. Appl. Genet.* **2004**, *108*, 401–413. [CrossRef] [PubMed]

150. Pozniak, C.J.; Knox, R.E.; Clarke, F.R.; Clarke, J.M. Identification of QTL and association of a phytoene synthase gene with endosperm colour in durum wheat. *Theor. Appl. Genet.* **2007**, *114*, 525–537. [CrossRef] [PubMed]

151. Peleg, Z.; Cakmak, I.; Ozturk, L.; Yazici, A.; Jun, Y.; Budak, H.; Saranga, Y. Quantitative trait loci conferring grain mineral nutrient concentrations in durum wheat×wild emmer wheat RIL population. *Theor. Appl. Genet.* **2009**, *119*, 353–369. [CrossRef] [PubMed]

152. Zhang, W.; Chao, S.; Manthey, F.; Chicaiza, O.; Brevis, J.C.; Echenique, V.; Dubcovsky, J. QTL analysis of pasta quality using a composite microsatellite and SNP map of durum wheat. *Theor. Appl. Genet.* **2008**, *117*, 1361–1377. [CrossRef] [PubMed]

153. Royo, C.; Soriano, J.M.; Alvaro, F. Wheat: A crop in the bottom of the Mediterranean diet pyramid. *Mediterr. Identities Environ. Soc. Cult.* **2017**, 381–399. [CrossRef]

154. Nachit, M.M.; Elouafi, I. Durum wheat adaptation in the Mediterranean dryland: Breeding, stress physiology, and molecular markers. In *Challenges and Strategies of Dryland Agriculture*; Rao, S.C., Ryan, J., Eds.; CSSA Special Publication: Madison, WI, USA, 2004; pp. 203–218.

155. Prat, N.; Guilbert, C.; Prah, U.; Wachter, E.; Steiner, B.; Langin, T.; Buerstmayr, H. QTL mapping of Fusarium head blight resistance in three related durum wheat populations. *Theor. Appl. Genet.* **2017**, *130*, 13–27. [CrossRef]

156. Varella, A.C.; Zhang, H.; Weaver, D.K.; Cook, J.P.; Hofland, M.L.; Lamb, P.; Talbert, L.E. A novel QTL in durum wheat for resistance to the wheat stem sawfly associated with early expression of stem solidness. *G3 Genes Genomes Genet.* **2019**, *9*, 1999–2006. [CrossRef]

157. Milner, S.G.; Maccaferri, M.; Huang, B.E.; Mantovani, P.; Massi, A.; Frascaroli, E.; Salvi, S. A multiparental cross population for mapping QTL for agronomic traits in durum wheat (*Triticum turgidum* ssp. *durum*). *Plant Biotechnol. J.* **2016**, *14*, 735–748. [CrossRef]

158. Peña, R.J. Wheat for bread and other foods. In *Bread Wheat Improvement and Production*; Curtis, B.C., Rajaram, S., Gómez Macpherson, H., Eds.; The Food and Agriculture Organization of the United Nations (FAO): Rome, Italy, 2002; pp. 483–542.

159. Gale, K.R. Diagnostic DNA markers for quality traits in wheat. *J. Cereal Sci.* **2005**, *41*, 181–192. [CrossRef]

160. Blanco, A.; Bellomo, M.P.; Lotti, C.; Maniglio, T.; Pasqualone, A.; Simeone, R.; Troccoli, A.; Fonzo, N.D. Genetic mapping of sedimentation volume across environments using recombinant inbred lines of durum wheat. *Plant Breed.* **1998**, *117*, 413–417. [CrossRef]

161. Ma, W.; Zhang, W.; Gale, K.R. Multiplex-PCR typing of high molecular weight glutenin alleles in wheat. *Euphytica* **2003**, *134*, 51–60. [CrossRef]

162. Yıldırım, A.; Sönmezoğlu, Ö.A.; Sayaslan, A.; Kandemir, N.; Gökmen, S. Molecular breeding of durum wheat cultivars for pasta quality. *Qual. Assur. Saf. Crop.* **2019**, *11*, 15–21. [CrossRef]

163. Blanco, A.; de Giovanni, C.; Laddomada, B.; Sciancalepore, A.; Simeone, R.; Devos, K.M.; Gale, M.D. Quantitative trait loci influencing grain protein content in tetraploid wheats. *Plant Breed.* **1996**, *115*, 310–316. [CrossRef]

164. Edwards, N.M.; Gianibelli, M.C.; McCaig, T.N.; Clarke, J.M.; Ames, N.P.; Larrique, O.R.; Dexter, G.E. Relationship between gluten strength, polymeric protein quantity and composition for diverse durum wheat genotypes g. *Cereal Sci.* **2007**, *45*, 140–149. [CrossRef]

165. Sissons, M. Role of durum wheat composition on the quality of pasta and bread. *Food* **2008**, *2*, 75–90.

166. Giuliani, M.M.; De Santis, M.A.; Pompa, M.; Giuzio, I.; Flagella, Z. Analysis of gluten proteins composition during grain filling in two durum wheat cultivars submitted to two water regoms. *Ital. J. Agron.* **2014**, *9*, 558.

167. Giuliani, M.M.; Palermo, C.; De Santis, M.A.; Mentana, A.; Pompa, M.; Giuzio, I.; Masci, S.; Centonze, D.; Flagella, Z. Differential expression of durum wheat glutenproteome under water stress during grain filling. *J. Agric. Food Chem.* **2015**, *63*, 6501–6512. [CrossRef]

168. D'Ovidio, R.; Masci, S. The wheat low-molecular-weight glutenin subunits. *J. Cereal Sci.* **2004**, *39*, 321–339. [CrossRef]

169. Sosinov, A.A.; Poperelya, F.A. Prolamin polymorphism and plant breeding. *Bull. Agric. Sci.* **1979**, *10*, 21–34.

170. Wrigley, C.W.; Autran, J.C.; Bushuk, W. Identification of cereal varieties by gel electrophoresis of grain proteins. *Adv. Cereal Sci. Technol.* **1982**, *5*, 211–259.

171. Melnikova, N.V.; Kudryavtseva, A.V.; Kudryatsev, A.M. Catalogue of alleles of gliadin-coding loci in durum wheat (*Triticum durum* Desg.). *Biochimie* **2011**, *94*, 551–557. [CrossRef] [PubMed]

172. Damideaux, R.; Autran, J.C.; Grignak, P.; Feillet, P. Mise en evidence de relations applicables en selection 921 entre l' electrophoregramme des gliadines et les properties viscoelastiques du gluten de *Triticum durum* 922 Desf. *C. R. Acad. Sci.* **1978**, *287*, 701–704.

173. Poperelya, F.A.; Babayants, I.G. Gliadin component block Gld-183 as a marker of gene determining stem rust resistance. *Dokl. Vaskhnil* **1978**, *6*, 6–7.

174. Raciti, C.N.; Doust, M.A.; Lombardo, G.M.; Boggini, G.; Pecetti, L. Characterization of durum wheat mediterranean germplasm for high and low molecular weight glutenin subunits in relation with quality. *Eur. J. Agron.* **2003**, *19*, 373–382. [CrossRef]

175. Rossini, F.; Provenzano, M.E.; Sestili, F.; Ruggeri, R. Synergistic effect of sulfur and nitrogen in the organic and mineral fertilization of durum wheat: grain yield and quality traits in the Mediterranean environment. *Agronomy* **2018**, *8*, 189. [CrossRef]

176. Li, L.; Niu, Y.; Ruan, Y.; DePauw, R.M.; Singh, A.K.; Gan, Y. Agronomic advancement in tillage, crop rotation, soil health, and genetic gain in durum wheat cultivation: A 17-year Canadian story. *Agronomy* **2018**, *8*, 193. [CrossRef]

177. Morris, C.E.; Sands, D.C. The breeder's dilemma-yield or nutrition? *Nat. Biotechnol.* **2006**, *24*, 1078–1080. [CrossRef]

178. Scherf, K.A. Immunoreactive cereal proteins in wheat allergy, non-celiac gluten/wheat sensitivity (NCGS) and celiac disease. *Curr. Opin. Food Sci.* **2019**, *25*, 35–41. [CrossRef]

179. Ziegler, K.; Neumann, J.; Liu, F.; Fröhlich-Nowoisky, J.; Cremer, C.; Saloga, J.; Reinmuth-Selzle, K.; Pöschl, U.; Schuppan, D.; Bellinghausen, I.; et al. Nitration of wheat amylase trypsin inhibitors increases their innate and adaptive immunostimulatory potential In Vitro. *Front. Immunol.* **2019**, *9*, 3174–3184. [CrossRef]

180. Cianferoni, A. Wheat allergy: Diagnosis and management. *J. Asthma Allergy* **2016**, *9*, 13–25. [CrossRef]

181. Korte, A.; Farlow, A. the advantages and limitations of trait analysis with gwas: A review. *Plant Methods* **2013**, *9*, 29. [CrossRef] [PubMed]

182. Mangini, G.; Gadaleta, A.; Colasuonno, P.; Marcotuli, I.; Signorile, A.M.; Simeone, R.; De Vita, P.; Mastrangelo, A.M.; Laidò, G.; Pecchioni, N.; et al. Genetic dissection of the relationships between grain yield components by genome-wide association mapping in a collection of tetraploid wheats. *PLoS ONE* **2018**, *13*, e0190162. [CrossRef] [PubMed]

183. Wang, S.; Xu, S.; Chao, S.; Sun, Q.; Liu, S.; Xia, G. A genome-wide association study of highly heritable agronomic traits in durum wheat. *Front. Plant Sci.* **2019**, *10*, 919. [CrossRef] [PubMed]

184. Sivakumar, S.; Matthew, P.R.; Sansaloni, C. Genome-Wide association analyses identify QTL hotspots for yield and component traits in durum wheat grown under yield potential, drought, and heat stress environments. *Front. Plant Sci.* **2018**, 81. [CrossRef]

185. Aoun, M.; Breiland, M.; Turner, K.M.; Loladze, A.; Chao, S.; Xu, S.; Acevedo, M. Genome-Wide association mapping of leaf rust response in a durum wheat worldwide germplasm collection. *Plant Genome* **2016**, *9*. [CrossRef]

186. Liu, W.; Maccaferri, M.; Bulli, P.; Rynearson, S.; Tuberosa, R.; Chen, X.; Pumphrey, M. Genome-Wide association mapping for seedling and field resistance to *Puccinia striiformis* f. sp. *tritici in elite durum wheat*. *Theor. Appl. Genet.* **2017**, *130*, 649–667. [CrossRef]

187. Liu, W.; Maccaferri, M.; Rynearson, S.; Letta, T.; Zegeye, H.; Tuberosa, R.; Pumphrey, M. Novel sources of stripe rust resistance identified by genome-wide association mapping in Ethiopian durum wheat (*Triticum turgidum* ssp. *durum*). *Front. Plant Sci.* **2017**, *8*, 774. [CrossRef]

188. Kidane, Y.G.; Hailemariam, B.N.; Mengistu, D.K.; Fadda, C.; Pè, M.E.; Dell'Acqua, M. Genome-Wide association study of septoria tritici blotch resistance in ethiopian durum wheat landraces. *Front. Plant Sci.* **2017**, *8*, 1586. [CrossRef]

189. Marcotuli, I.; Houston, K.; Waugh, R.; Fincher, G.B.; Burton, R.A.; Blanco, A.; Gadaleta, A. Genome wide association mapping for arabinoxylan content in a collection of tetraploid wheats. *PLoS ONE* **2015**, *10*, e0132787. [CrossRef]

190. Nigro, D.; Gadaleta, A.; Mangini, G.; Colasuonno, P.; Marcotuli, I.; Giancaspro, A.; Blanco, A. Candidate genes and genome-wide association study of grain protein content and protein deviation in durum wheat. *Planta* **2019**, *249*, 1157–1175. [CrossRef]

191. Wang, S.; Wong, D.; Forrest, K.; Allen, A.; Chao, S.; Huang, B.E.; Mastrangelo, A.M. Characterization of polyploid wheat genomic diversity using a high-density 90,000 single nucleotide polymorphism array. *Plant Biotechnol. J.* **2014**, *12*, 787–796. [CrossRef] [PubMed]

192. Avni, R.; Nave, M.; Eilam, T.; Sela, H.; Alekperov, C.; Peleg, Z.; Distelfeld, A. Ultra-Dense genetic map of durum wheat× wild emmer wheat developed using the 90K iSelect SNP genotyping assay. *Mol. Breed.* **2014**, *34*, 1549–1562. [CrossRef]

193. Maccaferri, M.; Ricci, A.; Salvi, S.; Milner, S.G.; Noli, E.; Martelli, P.L.; Ammar, K. A high-density, SNP-based consensus map of tetraploid wheat as a bridge to integrate durum and bread wheat genomics and breeding. *Plant Biotechnol. J.* **2015**, *13*, 648–663. [CrossRef] [PubMed]

194. Maccaferri, M.; Harris, N.S.; Twardziok, S.O.; Pasam, R.K.; Gundlach, H.; Spannagl, M.; Himmelbach, A. Durum wheat genome highlights past domestication signatures and future improvement targets. *Nature Genet.* **2019**, *51*, 885. [CrossRef]

195. Krasileva, K.V.; Buffalo, V.; Bailey, P.; Pearce, S.; Ayling, S.; Tabbita, F.; Dubcovsky, J. Separating homeologs by phasing in the tetraploid wheat transcriptome. *Genome Biol.* **2013**, *14*, R66. [CrossRef] [PubMed]

196. Vendramin, V.; Ormanbekova, D.; Scalabrin, S.; Scaglione, D.; Maccaferri, M.; Martelli, P.; Tuberosa, R. Genomic tools for durum wheat breeding: De novo assembly of Svevo transcriptome and SNP discovery in elite germplasm. *BMC Genom.* **2019**, *20*, 278. [CrossRef] [PubMed]

197. Krasileva, K.V.; Vasquez-Gross, H.A.; Howell, T.; Bailey, P.; Paraiso, F.; Clissold, L.; Fosker, C. Uncovering hidden variation in polyploid wheat. *Proc. Natl. Acad. Sci. USA* **2017**, *114*, 913–921. [CrossRef]

198. Sieber, A.N.; Longin, C.F.H.; Würschum, T. Molecular characterization of winter durum wheat (*Triticumdurum*) based on a genotyping-by-sequencing approach. *Plant Genet. Res.* **2017**, *15*, 36–44. [CrossRef]

199. Baloch, F.S.; Alsaleh, A.; Shahid, M.Q.; Çiftçi, V.; de Miera, L.E.S.; Aasim, M.; Hatipoğlu, R. A whole genome DArTseq and SNP analysis for genetic diversity assessment in durum wheat from central fertile crescent. *PLoS ONE* **2017**, *12*, e0167821. [CrossRef]

200. Newell, M.A.; Jannink, J.L. Genomic selection in plant breeding. In *Crop Breed*; Humana Press: New York, NY, USA, 2014; pp. 117–130.

201. Hayes, B.J.; Goddard, M.E. Prediction of total genetic value using genome-wide dense marker maps. *Genetics* **2001**, *157*, 1819–1829.

202. Fiedler, J.D.; Salsman, E.; Liu, Y.; Michalak de Jiménez, M.; Hegstad, J.B.; Chen, B.; Li, X. Genome-Wide association and prediction of grain and semolina quality traits in durum wheat breeding populations. *Plant Genome* **2017**, *10*. [CrossRef] [PubMed]

203. Watson, A.; Ghosh, S.; Williams, M.J.; Cuddy, W.S.; Simmonds, J.; Rey, M.D.; Adamski, N.M. Speed breeding is a powerful tool to accelerate crop research and breeding. *Nat. Plants* **2018**, *4*, 23. [CrossRef] [PubMed]

204. Alahmad, S.; Dinglasan, E.; Leung, K.M.; Riaz, A.; Derbal, N.; Voss-Fels, K.P.; Hickey, L.T. Speed breeding for multiple quantitative traits in durum wheat. *Plant Methods* **2018**, *14*, 36. [CrossRef] [PubMed]

205. Altpeter, F.; Vasil, V.; Srivastava, V.; Vasil, I.K. Integration and expression of the high-molecular-weight glutenin subunit 1Ax1 into wheat. *Nat. Biotechnol.* **1996**, *14*, 1155–1159. [CrossRef] [PubMed]

206. He, G.Y.; Rooke, L.; Steele, S.; Bekes, F.; Gras, P.; Tatham, A.S.; Fido, R.; Barcelo, P.; Shewry, P.R.; Lazzeri, P.A. Transformation of pasta wheat (*Triticum durum* L. var. *durum*) with high-molecular-weight glutenin subunit genes and modification of dough functionality. *Mol. Breed.* **1999**, *5*, 377–396. [CrossRef]

207. Wang, H.; Yang, H.; Shivalila, C.S.; Dawlaty, M.M.; Cheng, A.W.; Zhang, F.; Jaenisch, R. One-Step generation of mice carrying mutations in multiple genes by CRISPR/Cas-mediated genome engineering. *Cell* **2013**, *153*, 910–918. [CrossRef]

208. Shan, Q.; Wang, Y.; Li, J.; Gao, C. Genome editing in rice and wheat using the CRISPR/Cas system. *Nat. Protoc.* **2014**, *9*, 2395–2410. [CrossRef]

209. Gil-Humanes, J.; Wang, Y.; Liang, Z.; Shan, Q.; Ozuna, C.V.; Sánchez-León, S.; Baltes, N.J.; Starker, C.; Barro, F.; Gao, C.; et al. High-Efficiency gene targeting in hexaploid wheat using DNA replicons and CRISPR/Cas9. *Plant J.* **2017**, *89*, 1251–1262. [CrossRef]

210. Zhang, Y.; Liang, Z.; Zong, Y.; Wang, Y.; Liu, J.; Chen, K.; Qiu, J.-L.; Gao, C. Efficient and transgene-free genome editing in wheat through transient expression of CRISPR/Cas9 DNA or RNA. *Nat. Commun.* **2016**, *7*, 1–8. [CrossRef]

211. Uauy, C.; Paraiso, F.; Colasuonno, P.; Tran, R.K.; Tsai, H.; Berardi, S.; Comai, L.; Dubcovsky, J. A modified TILLING approach to detect induced mutations in tetraploid and hexaploid wheat. *BMC Plant Biol.* **2009**, *9*, 115. [CrossRef]

212. Varshney, R.K.; Sinha, P.; Singh, V.K.; Kumar, A.; Zhang, Q.; Bennetzen, J.L. 5Gs for crop genetic improvement. *Curr. Opin. Plant Biol.* **2020**. [CrossRef] [PubMed]

 *agronomy*

Review

Durum Wheat (*Triticum durum* Desf.): Origin, Cultivation and Potential Expansion in Sub-Saharan Africa

Amadou Tidiane Sall [1,2], Tiberio Chiari [3], Wasihun Legesse [4], Kemal Seid-Ahmed [2], Rodomiro Ortiz [5], Maarten van Ginkel [6] and Filippo Maria Bassi [2,*]

1   Institut Sénégalais de Recherches Agricoles (ISRA), Saint-Louis 46024, Senegal; tidianeassall11@gmail.com
2   International Center for Agricultural Research in the Dry Areas (ICARDA), Rabat 10000, Morocco; s.a.kemal@cgiar.org
3   Italian Agency for Development Cooperation (AICS), Addis Ababa 1000, Ethiopia; tiberio.chiari@aics.gov.it
4   Ethiopian Institute Agricultural Research (EIAR), Agricultural Research Center, Debre Zeit 1000, Ethiopia; wasihunl@yahoo.com
5   Sveriges lantbruksuniversitet (SLU), Institutionen för Växtförädling (VF), 23053 Alnarp, Sweden; rodomiro.ortiz@slu.se
6   International Center for Agricultural Research in the Dry Areas (ICARDA), Amman 1118, Jordan; maartenvanginkel1951@gmail.com
*   Correspondence: F.Bassi@cgiar.org; Tel.: +212-614-402-717

Received: 11 April 2019; Accepted: 13 May 2019; Published: 24 May 2019

Abstract: Durum wheat is an important food crop in the world and an endemic species of sub-Saharan Africa (SSA). In the highlands of Ethiopia and the oases of the Sahara this crop has been cultivated for thousands of years. Today, smallholder farmers still grow it on marginal lands to assure production for their own consumption. However, durum wheat is no longer just a staple crop for food security but has become a major cash crop. In fact, the pasta, burghul and couscous industry currently purchase durum grain at prices 10 to 20% higher than that of bread wheat. Africa as a whole imports over €4 billion per year of durum grain to provide the raw material for its food industry. Hence, African farmers could obtain a substantial share of this large market by turning their production to this crop. Here, the achievements of the durum breeding program of Ethiopia are revised to reveal a steep acceleration in variety release and adoption over the last decade. Furthermore, the variety release for Mauritania and Senegal is described to show how modern breeding methods could be used to deliver grain yields above 3 t ha^{-1} in seasons of just 92 days of length and in daytime temperatures always above 32 °C. This review describes the potential of releasing durum wheat varieties adapted to all growing conditions of SSA, from the oases of the Sahara to the highlands of Ethiopia. This indicates that the new breeding technologies offer great promise for expanding the area of durum wheat production in SSA but that this achievement remains primarily dependent on the market ability to purchase these grains at a higher price to stimulate farmer adoption. The critical importance of connecting all actors along the semolina value chain is presented in the example of Oromia, Ethiopia and that success story is then used to prompt a wider discussion on the potential of durum wheat as a crop for poverty reduction in Africa.

Keywords: Agro-industry; Ethiopia; oasis wheat; pasta wheat; Senegal River; value chain

---

1. Introduction

Durum wheat (*Triticum durum* Desf.) is an important food crop of the world, with an estimated 36 million t of annual global production [1]. The largest producing countries are Turkey and Canada with estimated 2 million ha each [2,3], followed by Algeria, Italy and India, each cultivating over

1.5 million ha [4–6]. Syria belonged to this group of large producers but the recent unrest has strongly reduced crop production. France, Greece, Morocco, Pakistan, Portugal, Kazakhstan, Russia, Spain and Tunisia cultivate durum wheat on between 0.5 and 0.8 million ha annually [3]. Azerbaijan, Iraq and Iran combined grow durum wheat on over 0.7 million ha [6]. In addition, Egypt, Jordan and Lebanon grow it on relatively large areas [7–9]. The Sonora desert and other small areas of Mexico also target the production of this crop for the export market on approximately 0.2 million ha [10]. Australia is similarly exploring the cultivation of this crop with 0.1 million ha allocated annually to its production [11]. In sub-Saharan Africa (SSA), Ethiopia is the largest producer of durum wheat, with approximately 0.6 million ha [12].

A very large amount of genetic diversity exists for this crop and that diversity also extends to the many traditional ways of consuming it, including several unique dishes that represent with pride the national identities: pasta, couscous, bourghul, *freekeh*, *gofio* and unleavened breads, just to name a few [13]. Regardless of its tight connection to the dishes of the tradition, durum wheat today is cultivated in developed countries mainly as a cash crop to feed the booming food industry. The annual production of pasta was estimated at 14.3 million t in 2013, with a global market approximated at €14.9 billion and average global price of 1045 € t^{-1} [14]. On a global scale, most of its consumption and production are in Europe, South America and the United States of America. Africa accounts only for 5.6% of total pasta production, mainly in Egypt, South Africa and Tunisia [14] and Asia consumption is also on the raise. Detailed data for SSA are hard to obtain, as most statistics combine durum wheat with bread wheat into single "wheat" data points but the estimations that could be gathered from several sources suggest an import market of €337 million and an export market mostly within the continent of €40 million (Table 1). Reliable data on the size of the internal market were not found. In this review, the developing couscous and bourghul industrial markets are not included, as data are not readily available. Italy, North African, South Africa and Turkey are the largest exporters of pasta to SSA [15]. However, the total area dedicated to durum wheat in SSA is limited to 630,000 ha, of which 90% is cultivated in Ethiopia. Therefore, this is the only country capable of producing pasta using locally grown grain, while for all other SSA countries the bulk of pasta production required the import of €483 million worth of durum grain from Canada, Turkey and the USA (Table 1). It must be mentioned that the pasta industry in SSA often utilizes bread wheat flour for its production and typically only products from North Africa and developed countries meet the international standard definition of 'pasta' by using 100% durum semolina [14]. Clearly, there is huge agricultural and commercial scope for expanding domestic production and marketing of durum wheat in SSA countries.

Durum wheat and rice are the most lucrative among the cereals, with prices usually 20 to 40% higher than common wheat, millet, maize and sorghum [16]. While durum wheat remains a critical staple food for smallholder farmers in marginal lands, thanks to its exceptional adaptation to climatic stresses, its large-scale production is tightly linked to its greater monetary return. In the absence of governmental subsidies that push toward the cultivation of other crops, farmers tend to prefer durum wheat as long as the market continues to guarantee additional profits. In this regard, the existence of a strong value chain for the pasta, couscous and bourghul industry is quintessential to the success of durum wheat cropping.

Table 1. Economy and production of durum wheat in Africa and sub-Saharan Africa.

Country	Crop Use per Land			Durum Exports [b]		Durum Imports [b]	
	Rice [a]	Wheat (all) [a]	Durum	Grains	Pasta	Grains	Pasta
	(ha)	(ha)	(ha)	(€)	(€)	(€)	(€)
East	**74,069**	**1,931,714**	**599,552**	**365,889**	**247,159**	**129,119,247**	**46,339,452**
Eritrea	-	25,000	12,500 [c]	-	-	-	8,210,000
Ethiopia	33,820	1,605,654	561,979 [d]	-	28,079	129,113,334 [n]	24,433,488 [n]
Kenya	31,349	162,900	16,290 [e]	365,889	219,080	2177	13,695,964
Somalia	1338	2500	na	-	-	3735	-
Sudan	7562	135,660	6783 [f]	-	-	na	na
Central	**206,592**	**44,687**			**1,636,218**	**11,864,150**	**21,908,766**
Burundi	21,670	8828	na	-	1324	3136	226,183
Cameroon	166,734	660	na	-	1,487,086	11,858,665 [n]	13,442,925 [n]
Gabon	620	-	-	-	69,532	2349	5,665,685
Rwanda	17,568	35,199	na	-	78,276	-	2,573,974
South	**1,437,257**	**608,622**	**26,521**	**38,590,262**	**11,794,879**	**148,416,094**	**86,270,994**
Angola	29,510	3420	720 [g]	-	-	1229	-
Madagascar	909,492	2087	na	-	2730	142,521	26,445,917
Malawi	65,275	1269	na	-	3862	34,292,325 [n]	808,085
Mozambique	300,000	18,081	-	1295	-	-	3,568,930
South Africa	1150	505,500	23,456 [h]	35,183,511	11,702,825	33,889,190	31,108,885
Uganda [1]	93,000	14,000	100 [i]	86,843	76,540	3,673,000	3,935,998
Zambia	38,520	41,810	0 [j]	3,318,613	2177	-	5,982,422
Zimbabwe	310	22,455	2246 [k]	-	6745	76,417,829	14,420,757
West	**7,394,599**	**102,033**	**4302**	**1,123,513**	**87,653,612**	**193,535,341**	**155,068,054**
Benin	68,259	-		474,234	222,295	8,365,599 [n]	50,781,356 [n]
Burkina Faso	138,852	-	-	-	65,747	272,228	22,915,826 [n]
Gambia	66,380	-	-	-	536,030	-	5,094,377
Ghana	215,905	-	-	2340	3,633,396	78,376	7,847,000 [o]
Guinea	1,642,687	-	-	-	111,862	17,673,754	3,780,845
Ivory Coast	791,691	-	-	646,939	32,545,382	28,513,879 [o]	3,024,576 [o]
Mali	604,745	9947	1100 [k]	-	211,989	-	16,050,025
Mauritania	43,900	1700	na [l]	-	19,077	51,161,839	14,377,271
Niger	21,572	1883	na	-	7,328,976	3,101,758	20,659,562 [n]
Nigeria	2,931,400	80,000	na	-	40,917,371	38,300,083 [n]	2,633,648
Senegal	108,547	3	2 [m]	-	2,061,487	46,058,643 [n]	7,903,568
Sierra Leone	671,422	-	-	-	-	9182	-
Togo	92,239	-	-	-	2,099,984	-	27,284,750 [n]
sub-Saharan Africa	**9,112,517**	**2,687,056**	**632,375**	**40,079,663**	**101,331,868**	**482,934,832**	**309,587,266**
Africa	**9,714,796**	**9,960,981**	**3,557,740**	**40,346,664**	**215,258,164**	**4,166,972,506**	**375,508,531**

[1] The former Sudan is not a true Sub-Saharan country, but it has agro-environmental conditions that differ from North Africa and therefore it is presented here. Data consider Sudan and South Sudan together. [2] Uganda is reported among Southern Africa countries instead of East Africa for its closer similarity in the use of durum wheat. [a] Data obtained from FAOSTAT for the 2013 season [17]. [b] Data obtained from The Economic Complexity Observatory of the year 2013 [15], except when otherwise indicated. [c] [18]. [d] [19]. [e] [20] [f] [21]. [g] [22]. [h] [23]. [i] [24]. [j] [25]. [k] [26]. [l] [17]. [m] Land surface utilized on-station. [n] Data confirmed on Fact-Fish [27]. [o] Data confirmed on Index Mundi [28]. The **bold** rows indicate sub-totals for different African regions.

In this review, the current status of durum wheat production in SSA is discussed in comparison to the needs of the local pasta industry to better understand the potential of its expansion through the deployment of novel adapted varieties. Because of its industrial nature, durum wheat has often been disregarded by SSA policy makers in favour of bread wheat as a more direct "food security" approach. However, among the sustainable development goals set by the United Nations, "poverty reduction" is considered as a strategic way to tackle famine, without causing nutritional deficits due to mono-food diets. In this sense, durum wheat is at least as well suited as bread wheat in improving livelihoods. Both aspects of durum wheat, as a "food security" staple food for smallholder farmers, as well as a "poverty reduction" industrial crop will be considered here.

2. An Endemic Crop of SSA: Durum Wheat Second Centre of Origin in Ethiopia

Durum wheat originated from the domesticated form of a wild species named emmer wheat (*Triticum dicoccum* Koern.) between 12,000 and 10,000 years ago, in the West Levantine [29]. Phoenicians have traded it along the Mediterranean shores since historical times and throughout the rise of civilizations this crop has encountered several waves of expansion until today's global importance [30]. However, durum wheat did not originate solely in West Asia. Archaeological evidence suggests that naked emmer reached Ethiopia approximately 5000 years ago [31], probably arriving from the Levantine, through Egypt, along the Silk Road [32]. Today emmer wheat occupies approximately 7% of the wheat production in Ethiopia under the local name of *aja*. Recent molecular data [33] indicated that Ethiopian farmers repeated what had been achieved already in West Asia before, by deriving durum wheat anew through the further domestication of emmer. This new origin of the same crop gave rise to a subspecies known as *T. turgidum* ssp. *aethiopicum* or *abyssinicum*. Until relatively recently, landraces belonging to this subspecies were widely cultivated by smallholder farmers in Ethiopia, with up to 80% of the total durum land farmed with these unique biotypes [34]. The highlands of Ethiopia are known areas of rich biodiversity and durum wheat is no exception [35,36]. For instance, one of the unique characteristics identified among *T. aethiopicum* landraces is the purple colour of the grains, particularly rich in anthocyanins [37]. Anthocyanins act as anti-oxidants and provide other health benefits, hence these could be potentially exploited by the pasta industry to develop extra nutritious food products. Morphological and molecular characterization of these landraces has only just begun and already several traits such as resistance to diseases (e.g., stem rust, powdery mildew), drought tolerance, long coleoptile, high tillering and resistance sources to Hessian fly have been identified [38,39]. This biodiversity has already started attracting strong interest by the international community for utilization, pushing the Ethiopian Government to protect it under strict germplasm exchange policy [18]. In order to conserve these resources, the Ethiopian Biodiversity Institute (EBI) has established a holding of over 7000 accessions collected from different parts of Ethiopia [34]. These collections have been extensively investigated for their morphological and molecular diversity by many researchers and useful traits were identified and are now utilized by breeders and plant genetic conservationists in Ethiopia and beyond [40–48]. In the past two decades, the acreage of traditional tetraploid wheat has drastically diminished due to displacement by improved bread wheat varieties, extensive cultivation of *Tef* and Kabuli chickpea, farmland fragmentation, policies favouring bread wheat and the absence of a strong seed supply system [49]. To reduce this genetic erosion, EBI has established in situ conservation sites to conserve the agro-biodiversity at the farm level in different parts of Ethiopia. Community biodiversity practices were established in East Shoa and South Wollo zones with the aim of establishing community seed banks, participatory variety selection and the re-introduction of local durum wheat biotypes, food legumes and sorghum into the cropping system [50,51]. Regardless of their specific uses, these landraces represent a treasure chest of potentially new and useful traits that breeders could be able to exploit to deliver superior varieties with added market values.

3. Durum Wheat in East Africa as a Staple and Cash Crop

East African countries cultivate almost 2 million ha of wheat, of which only 630,000 ha are farmed with durum wheat (Table 2). Eritrea, Kenya, Somalia and Sudan combined harvested as little as 37,000 ha of durum wheat in 2014. Yet, these countries have maintained in their culinary taste the influence of the past Italian presence in the region, with pasta imports reaching 40 million USD in 2017 in Ethiopia only. In the case of Kenya, national production is sufficient to support the export of €0.5 million worth of pasta and durum grains.

Table 2. Durum wheat varieties currently cultivated in Sub-Saharan Africa.

Country	Variety Name	Adoption	Pedigree	Origin
Ethiopia	'Cocorit71'	Old variety, still cultivated	Enano/4*Tehuacan60// Stewart63/3/Anhinga	CIMMYT
Ethiopia	'Langdon(LD)357'	Old variety, still cultivated	LD308/Nugget	USA
Ethiopia	'Gerardo'	Old variety, still cultivated	GerardoVZ466/3/ ND61130/Leeds//Grulla	CIMMYT
Ethiopia	'Ejersa'	Variety utilized by farmers in Oromia	Labud/Nigris3// Gan	CIMMYT
Ethiopia	'Bakalcha'	Widely cultivated variety, now replaced due to susceptibility to stem rust	Gedirfa/Gwerou15	CIMMYT
Ethiopia	'Ude'	Variety that replaced Bakalcha in most zones	Chen/Altar// Jori69	CIMMYT
Ethiopia	'Mangudo'	Covers several districts in Oromia	Omruf1/Stojocri2/3/1718/ BeadWheat24//Karim	ICARDA
Ethiopia	'Asasa'	Low moisture area in Rift Valley	Cho/Taurus//Yav/3/Fg/4/ Cra/5/Fg/Dom/6/Hui	national
Ethiopia	'Utuba'	New favorite by farmers, cultivated already on 10,000 ha	Omruf1/Stojocri2/3/ 1718/BeadWheat24//Karim	ICARDA
Ethiopia	'Sinana1'	18,000 ha	Emmer selection from landraces	national
Ethiopia	'Lemesso'	18,000 ha	Emmer selection from landraces	national
Mauritania	'Karim'	Cultivated by farmers along the Senegal river and in oasis	Jori/Anhinga//Flamingo	CIMMYT
Mauritania, Senegal	'Haby'	New release under fats-track multiplication	Mrb5/T.dico Aleppo Col//Cham1	ICARDA
Mauritania, Senegal	'Elwaha'	New release under fats-track multiplication	Oslks/5/Azn/4/BezHF/3/ SD19539//Cham1/Gdr2	ICARDA
Mauritania, Senegal	'Bani Suef 5'	New release under fats-track multiplication	Dipperz/Bushen3	CIMMYT
Senegal	'Amina'	New release under fats-track multiplication	Korifla/AegSpeltoidesSyr// Loukos	ICARDA
Mali	'Biskri-Bouteille'	Old variety, still cultivated. Only available recorded release	Biskri/Bouteille	national
South Africa	'Kronos'	Most cultivated variety	APB MSFRS pop selection	USA
Kenya	'Mwewe'	Old variety, still cultivated	Flamingo/Leads	CIMMYT
Sudan	'Sham1'	Old variety, still cultivated	Plc/Ruff// Gta/Rtte	CIMMYT ICARDA
Eritrea	'Mindum XA10'	Old variety, still cultivated	Mindum/Asmara10	USA
Nigeria	'Anser8'	Holds potential for adoption	Altar84/Alondra//Sula	CIMMYT

The durum varieties used for production are old bred-lines from Centro Internacional de Mejoramiento de Maíz y Trigo (CIMMYT) and International Centre for Agricultural Research in the Dry Areas (ICARDA) such as 'Mwewe' (Flamingo/Leads), Mindum XA10 (Mindum/Asmara 10) and Sham 1 (Plc/Ruff//Gta/Rtte), in Eritrea, Kenya and Sudan, respectively (Table 2). The most critical traits of these varieties are earliness and tolerance to heat in irrigated Sudan and resistance to rust diseases under rainfed cultivation in Eritrea and Kenya. Information from Somalia is scarce and hard to obtain. Considering that the most cultivated durum varieties listed above are more than 30 years old, there is a significant genetic yield gap that could be filled through the release and commercialization of more modern varieties.

The Ethiopian case is presented in some detail, including critical historical steps, as it provides valuable lessons for other SSA countries planning to grow their durum wheat sector. In Ethiopia, durum wheat is produced predominantly in the Gojam, Gonder, Shewa, Tigrai and Wollo regions [52]. The main growers are smallholder farmers in the highlands, where the environmental characteristics are relatively low temperatures and high rainfall on black swelling/shrinking vertisol soils, with water logging as a common problem. The crop is planted late in the growing season to avoid early water logging and it continues to grow during the dry period on residual moisture at altitudes between 1800 and 2800 m.a.s.l. [44]. Due to late planting, it forfeits some of its additional potential yield in favour of higher protein content. The crop is consumed in several different forms such as unleavened breads, pancakes, macaroni, spaghetti, biscuits and pastries. The most common of the Ethiopian and Eritrean recipes include *dabo* (Ethiopian home-made bread), *hambasha* (bread from northern Ethiopia), *kitta* (unleavened bread), *injera* (thin bread normally made with *Tef*), *nifro* (boiled whole grains), *kolo* (roasted whole grains), *dabo kolo* (round and seasoned dough) and *kinche* (crushed kernels, cooked with milk or water and mixed with spiced butter). Besides the role of grain in traditional food and processed products, durum wheat straw is also greatly appreciated for its high palatability for livestock in the mixed farming systems of the highlands of Ethiopia [53]. Ethiopia today cultivates 562,000 ha of durum wheat [12], accounting for the vast majority of the cultivation of this crop in SSA (Table 2). Still, today's value represents just half of the land that was dedicated to durum wheat in 1967 [54] and this reduction continues in favour of more extensive farming of bread wheat [23]. This is the combined result of political will, the introduction of modern bread wheat cultivars that have replaced the traditional durum wheat landraces and the absence until now of vocal local industry demand of high quality pasta made from durum semolina. Ethiopia's push toward bread self-sufficiency has resulted in a monoculture of bread wheat (as well as maize), often cultivated in both the long (*meher*) and short (*belg*) rainy seasons, which in turn created a favourable environment of continuous host presence for the spread of damaging rust diseases and for the surge of tenacious weeds [55,56]. Tef, the largest cultivated crop in Ethiopia, also contributes to an expansion of monoculture in Ethiopian agriculture.

Durum wheat research in Ethiopia started back in 1949 at the Paradiso Experimental Station near Asmara [57]. Among several local durum landrace collections tested for productivity and stem and leaf rust resistance, four selections (A10, H23, P20 and R18) were developed and released to farmers in Eritrea in 1952. In 1956 and 1957, several crosses were made between local and exotic varieties mainly for the purpose of transferring the stem and leaf rust resistance of A10 and R18 to cv. 'Mindum' from the USA (Table 2). This resulted in two new varieties, which unfortunately had to be rapidly retracted due to susceptibility to new leaf rust races [58]. In the 1980s, the wheat research activities at the Paradiso station were discontinued and durum wheat breeding was transferred to the Debre Zeit Agricultural Research Centre [42]. At the Centre, many cultivars were developed and released, derived from landrace selections, local crosses and introductions from the international durum wheat breeding programs at CIMMYT and ICARDA. For clarity, in this review the word 'cultivar' has been used to define germplasm cultivated on large amounts of land, while the word 'variety' is reserved to define germplasm officially registered in the variety catalogue of one country. The first durum cultivars released from local breeding selections were 'Arendeto' (DZ04-118) and 'Marou' (DZ04-688), obtained by mass selection [59]. These were followed by the varieties 'Cocorit-71', 'LD-357' and 'Gerardo' obtained from the international agricultural research centres. Since 1982, a formal variety release system has been put in place, which also rationalized the previous work into a variety catalogue, which accounts today for 40 durum wheat cultivars (Figure 1). In the last two decades, many federal and regional agricultural research centres have become involved in durum wheat improvement to respond to the demand by 300 local flour and pasta manufacturers as well as the local consumers. This push by the national food industry, combined with a stronger presence in the region of international development agencies involved in breeding against the emerging Ug99 stem rust race threat [60], has resulted in an increase in the release of durum cultivars, with 20 varieties inscribed in the last 10 years [61]. These new varieties are more responsive to chemical inputs, resistant to diseases and can reach average yields of 4–5 t ha^{-1} under rainfed conditions [62]. 'Utuba' was released in 2015 as an alternative variety to 'Mangudo' and 'Mukuye' because of its amber seeds, high protein content and high yield potential. The grain yield performance on research station ranged from 3.4 to 6.5 t ha^{-1} and from 2.5 to 4.5 t ha^{-1} in farmers' fields [63]. 'Utuba' (Omruf1/Stojocri2/3/1718/BeadWheat24//Karim) takes on average 62 days to flower, 108 days to mature and it is also appreciated for its height (80–90 cm), which ensures good amount of straw for the livestock. A survey conducted by ICARDA has indicated that farmers that abandon the widely-grown durum cultivar 'Ude' (Chen/Altar//Jori69) to grow the recent release 'Utuba' (Table 2), obtain an average yield gain of 32% and an equivalent monetary return. Regardless of this clear advantage, adoption by farmers remains very low [64], mainly because of the high cost of purchasing quality seeds, scarce access to agriculture micro-credits and a national seed system incapable of reaching the more remote areas [65]. To solve some of these issues, international agricultural research centres and development agencies together with the national agricultural research institutes have launched a project to develop informal "Community Based Seed Enterprises" [66–69]. This informal system promotes farmers' aggregation around the possibility to gain access to improved seeds from their neighbours. Lead farmers are designated and provided free-of-charge with certified seeds of improved varieties. These leaders are then responsible for multiplying the seeds and providing them to their neighbours for a reasonable price agreed among each other, often involving exchange of livestock, land rental or payments after harvest. A significant effort has been made to expand the production of improved durum wheat cultivars to supply raw materials to the food industries. For example, in the 2018–2019 cropping season, the Bale Zone Bureau of Agriculture up-scaled the cultivation area of two durum wheat cultivars (cvs. Utuba and Mangudo) in nine districts covering over 6244 ha. In the north Shoa-Amhara region, the Africa RISING project, in partnership with the North Shoa Zone Bureau of Agriculture, expanded the area under these two cultivars to over 700 ha. This fast adoption pace is due to the national and international effort of promoting the new varieties but also the great farmers appreciation. Further, the recent contractual agreement between Minjar farmers and the ALVIMA pasta processing factory is predicted to provide an additional push to its adoption.

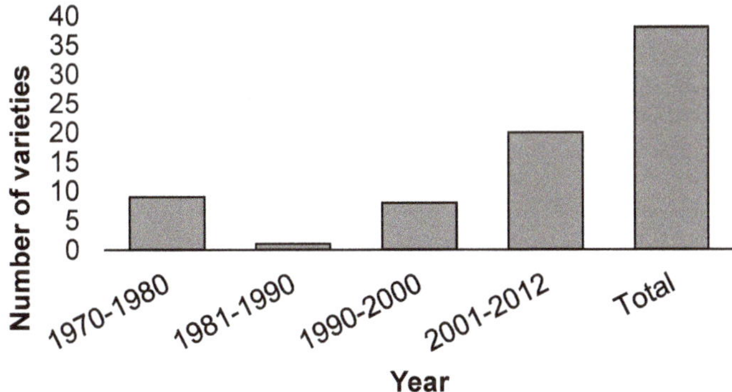

Figure 1. Durum variety releases in Ethiopia since 1970–2012.

Until today, Ethiopia still cultivates emmer wheat, the ancestor of durum wheat. Its cultivation is mainly restricted to marginal areas by about 300,000 households, covering 36,000 ha with an average productivity of 1.7 t ha^{-1} as recorded during the 2013–2014 season [70,71]. This area also continues to be drastically reduced due to expansion of modern bread wheat cultivars. Improvement of emmer wheat is given little attention and only two cultivars ('Sinana-1' and 'Lemesso') have been released through selection from landraces [61]. This crop is mainly used for the preparation of local food products such as *defo* or *dabo* (bread), *injera* (flat pancake bread), porridge, *kita* (flat steamed bread), *Kinche* (boiled coarse grain) and local drinks [72]. Emmer wheat is recommended for mothers as a special diet to maintaining their health and strength after childbirth because of its high protein content and digestibility [73]. In fact, its grain protein content ranges from 8.5 to 21.5%, which is 5–35% higher than in grain from oats or barley and it has a very low glycaemic index [74]. Emmer wheat is also a good source of resistance to leaf and stem rusts, powdery mildew, *Septoria* glume blotch, *Fusarium* head blight, Russian Wheat Aphid, in order of importance and tolerance to drought and heat [75–79].

4. Durum Wheat Value Chain in Oromia Region, Ethiopia

Recent investments in the pasta industry are proving extremely promising in Ethiopia thanks to new food habits of the growing urban populations, which are looking for fast and tasty foods, while still cheap and nutritious. Pasta has represented a ready-to-use option since its first introduction in Ethiopia in 1938 by the pioneering Italian enterprise Colonalpi (currently called Kaliti Food Share Company), later followed by the establishment of state-owned industries. Today the state industries have been privatized and grouped, together with numerous others, as members of the Ethiopian Millers Association. These pasta producers used to rely on massive importation of durum wheat grains, which was not a sustainable long-term business strategy due to high and volatile costs. Further, the purchase of foreign grains competed with other national priorities for the use of governmental hard currency stocks. Indeed, the revamping of national durum wheat production has caused the reduction of imports to negligible amounts in 2015 [80], after having equalled €129 million in 2013 (Table 1). However, at the same time, pasta import increased two-fold between 2011 and 2015, when it reached 50,000 t at a cost of about €40 million [80]. To reverse this trend, the Ethiopian Millers Association has eagerly explored the possibility to procure the needed raw material directly from local farmers to reduce production costs and increase competitiveness against foreign pasta imports. Unfortunately, the local production did not guarantee sufficient rheological grain quality to satisfy the industrial needs. In fact, grain of tetraploid landraces does not meet industrial standards in terms of colour or protein quality, while the high-yielding modern varieties tend to produce bleached and 'chalky' grains when grown on waterlogged vertisols in the absence of abundant nitrogen fertilization [81]. Hence, specific incentives needed to be provided to farmers to obtain industrial-grade harvests. The scope of the Ethiopian-Italian cooperation project for the Agricultural Value Chain in Oromia (AVCPO) was to re-direct some of the already existing bread wheat production system of the Bale zone toward the more lucrative farming of durum wheat for the industry. The process acted on the key elements required by the pasta industry to stabilize and self-sustain the value chain: competitive price, high rheological quality for conversion into pasta, easy and timely delivery, consistent stock of grains and predictable increases over years (Figure 2). Launched in April 2011, the initial steps relied on just two durum varieties (Table 2), identified as highly productive, resistant to prevailing diseases in the Bale zone and with good gluten strength: 'Ejersa' (Labud/Nigris3//Gan) and 'Bakalcha' (980SN Gedirfa/Gwerou15). A total of 40 t of certified seed were purchased from the Sinana Agricultural Research Centre (SARC). The dialogue with the pasta industries resulted in the signing of an innovative supply contract that set the purchase value to the prevailing bread wheat price, with the addition of a 'premium' strictly proportional to kernel protein content. This contract provided the needed incentive to farmers for the application of adequate fertilization strategies and has ensured high grain quality since. Furthermore, to supply the industry with large and uniform stocks of grains, AVCPO promoted farmer aggregation into 15 cooperatives and four unions and provided each with warehouses for temporary storage of grain. To measure the required quality, AVCPO equipped the SARC durum quality laboratory and trained researchers and technicians. Small-holder farmers cultivating around 0.5 to 2.0 ha of land were able to deliver their small sales to the warehouses and from there the industry could purchase large bulked stocks, as needed. Technical assistance to farmers and needed continuous research efforts were delivered by regional research and development institutions both from central and district-commune branches.

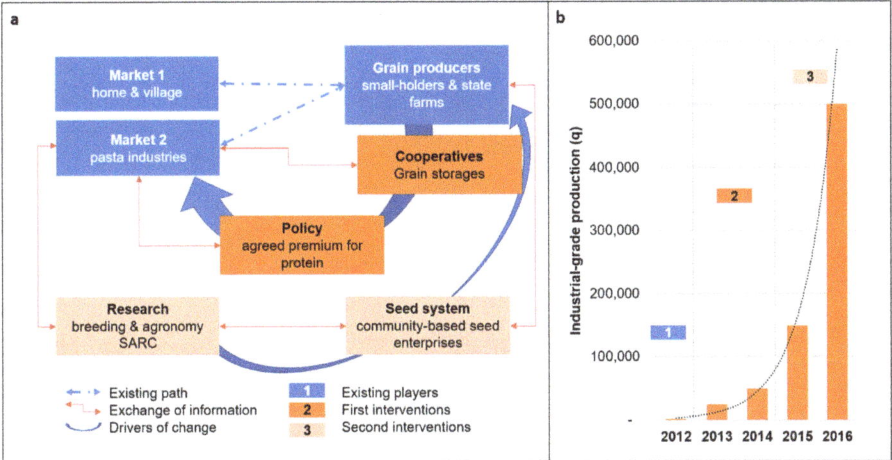

Figure 2. Durum wheat value chain in Oromia, Ethiopia. (**a**) Schematic of the intervention and value chain key actor relationships: to promote increased income in rural areas, the industrial requirements (market 2) were recorded and used to first promote contractual agreements for the sale of industrial-grade seeds and to assemble the farmers into cooperatives around grain storages and second to drive the research agenda with the release of superior cultivars and their multiplication via community-based seed enterprises; (**b**) Success indicator measured as the amount of durum grain sold to the food industry from the Oromia region since the inception of the project. SARC refers to the Sinana Agricultural Research Centre.

Highly innovative contractual relationships were created among farmer cooperatives and industries, pushing the surrounding authorities and public institutions to provide support and surveillance on proper accomplishment of duties. Among these, SARC formally acted as neutral third party for measuring the protein content and determining the final price. The emphasis on the highest level of participation and ownership by all involved stakeholders was considered as the key element for the success and sustainability of the development process [82]. Since the first harvest, durum wheat provided to farmers a significant monetary gain per ha of 25 to 30% over concurrent bread wheat and the industries were greatly satisfied with good rheological quality and reduced prices over imports. The availability of seed stocks of the two selected varieties enabled for prompt expansion of area planted through newly adopting farmers and cooperatives. Over time, the self-sufficient nature of the AVCPO's complex of cooperatives and institutions has created the premises for a vibrant market-oriented community eager to absorb and valorise new varieties and technologies developed by their research partners. Especially in the current situation of evolving rust races dramatically affecting bread wheat in the Bale and other wheat belts, farmers attribute to durum wheat the role of a rescue crop. By the convergence of all these factors, durum wheat production has exponentially increased from 500 t in 2011–2012, to a record harvest of 4.6 million t in 2017 due mainly to the 'Utuba' recent release and cultivation on large scale. 'Utuba' was christened and released as Ethiopian durum wheat variety in 2015 [83]. In the meantime, the value chain is already expanding to nearby Arsi and Shewa zones. The example of Oromia can be considered a successful approach on integration of the whole durum wheat value chain [84], with a proven rapid and sustainable impact. Hence, it provides a good example to follow for other SSA countries that rely today on durum wheat and pasta imports.

5. Durum Wheat in West Africa as a Future Cash Crop

West African countries cultivate over 7 million ha of irrigated rice but only 100,000 ha of wheat (mainly bread wheat) and mostly in Nigeria. A recent steep increase of wheat area has been reported for Nigeria but these data are not yet available from FAO statistics, the main source used for compiling Table 1. Still, all West African countries are importers of wheat grain and its derived products. A total of €155 million worth of pasta and €193 million worth of durum grains were imported in 2013 (Table 1). Benin is the largest importer of pasta in West Africa with almost €51 million worth imported in 2013, followed by Niger, Burkina Faso and Togo, which are also among the largest importers in Africa with €20, €22 and €27 million worth, respectively. Interestingly, €87 million worth of pasta are re-exported each year, mostly by Côte d' Ivoire and Nigeria. Since national durum production is close to zero, it means that large quantities of durum wheat grain are imported internationally, converted by the local industry into pasta products and then sold locally and to neighbouring countries. Hence, as was the case for the Oromia region in Ethiopia, there is potential for national durum cultivation to support this strong local industry, while sharing the €180 million worth per year of the current import market with the local growers. In Nigeria, initial steps have already been undertaken to identify suitable durum varieties at the Kadawa Kano field station. Here, 12 candidate varieties from CIMMYT's breeding program were assessed over two seasons. Trials revealed that grain yields exceeding 6.2 t ha^{-1} could be achieved in 100 days by the top performer 'Anser8' (Altar84/Alondra//Sula) under gravity irrigation [85].

Mauritania is the largest importer of durum grain in West Africa with over €51 million spent every year. This country has one of the most challenging agro-environments in West Africa, with farming substantially restricted to the narrow band along the Senegal River, where rainfall of up to 600 mm per year and irrigation water from the river sustain crop production (Agriculture in Mauritania, 2009). The Senegal River basin has a potential of irrigating 135,000 ha [86], of which less than 20% are currently utilized. The main crops are rice, pearl millet and cowpea. Wheat cultivation along the river is estimated at 8200 ha, of which approximately 5000 ha are grown with durum wheat. The only cultivated durum variety is 'Karim' (syn: 'Yavaros79', Jori/Anhinga//Flamingo), a widely adapted +35 years old CIMMYT-derived variety. Wheat is cultivated during the winter season in rotation with rice and cowpea under gravity irrigation. The window for growing wheat is rather narrow to avoid interfering with the cultivation of the two seasons of rice. Sowing has to occur between the end of November and the middle of December. The harvest is just 80 to 100 days later in early March. Regardless of this short season, two recent projects carried on at the experimental stations of Daara and Kaedi (U-Forsk2013 and SARD-SC) have revealed that yields of 3 t ha^{-1} could be reached along the Senegal River Valley. In response to these results, three new durum wheat varieties ('Haby' [Mrb5/T.dico Aleppo Col//Cham1], 'Elwaha' [Oslks/5/Azn/4/BezHF/3/SD19539//Cham1/Gdr2] and 'Bezater' [Ossl1/Stj5/5/Bicrederaa1/4/BEZAIZSHF//SD19539/Waha/3/Stj/Mrb3/6/Stj3//Bcr/Lks4/3/Ter3]) were released in 2016 (Table 2) and their seed multiplication has begun [87]. On the opposite shore of the river, the field station of Fanaye in Senegal obtained yields as high as 6 t ha^{-1}, when early planting towards the end of November was achieved. The irrigable agricultural land of Senegal is divided along three rivers, (in order of importance): Senegal, Faleme and Casamance, thus providing a total estimated irrigable land of 350,000 ha [88]. The Senegal River valley alone accounts for 240,000 ha of potential arable land [89], of which 110,000 ha are currently used for rice cultivation. Since 2017 small-holder farmers started growing improved heat tolerant durum wheat varieties after completion of the rice harvest, during the winter season, which is typically left at fallow. This research achievement, if sustained by polices and market demand, could help replace the €46 million worth of annual durum import by the national pasta industry. Furthermore, if the total rice area was to be converted to durum wheat instead of the fallow period, then this would be sufficient to generate an overproduction of durum grains to be exported to neighbouring countries for an interesting price. Just as the wheat-rice rotation system has been the cornerstone of India's food self-sufficiency with over 10 million ha still cultivated today [90], it can also become a new boost for the West African agriculture. In addition, the integration of a legume crop in the rotation with durum wheat and rice would be desirable to also increase long-term soil health and agro-ecosystem stability. In this regard, a suggestion is made to replace one rice season with cowpea, an excellent source of food and feed, with very high market value. The cropping model suggested would then become rice-durum wheat-cowpea. This expansion into considering a pulse such as cowpea as part of the durum wheat production system is, however, beyond the scope of this review and shall not be discussed further.

A third country relying on the Senegal River for irrigation is Mali, whose production is concentrated along this and the Niger River. The total irrigable land is estimated at 340,000 ha [91] with a potential to further expand. The vast majority of the land is utilized for the production of rice and maize during the warm months. Wheat is cultivated during winter on just 10,000 ha, of which a very small portion is durum wheat (Table 1). The old variety 'Biskri-Bouteille' (Biskri/Bouteille) is the only reported release for Mali [92]. It is likely that the breeding activities and import from neighbouring countries have resulted in more modern releases but no document could be located. Similar to its neighbours, Mali imports large quantities of pasta (€14 million) and part of it is further exported (€0.2 million). Hence, local production of durum wheat is a viable option for all three countries along the Senegal River. Their total area currently cultivated with rice reaches 754,000 ha. Assuming the same conditions apply to the whole surface, cultivation of durum wheat during the short winter fallow season has the potential to generate additional food, without reducing the current production of their main staple food. The newly identified super-early and heat tolerant durum varieties released in Mauritania and Senegal ('Haby', 'Elwaha', 'Bezater' and 'Amina': Korifla/AegSpeltoidesSyr//Loukos) can provide good industrial grain for the national industry and hold the potential to generate more than 1 million t of additional food in Sub-Saharan Africa [87].

The situation in Nigeria is no different than that observed for the Senegal River countries, even though, with over 80,000 ha farmed to wheat in 2013, it is already the largest bread wheat producer in West Africa (Table 1). A recent push by the Nigerian government, such as the removal of subsidies for the imported grains, has incentivized farmers to increase their wheat production and 2018/19 area harvested and production are estimated at 60,000 hectares and 60,000 tons, respectively [93]. Wheat is typically planted in November or December and harvested around April. The land used for wheat production is then rotated for other rainfed crops during the rainy season, which lasts in northern Nigeria from April to September. Rice is sometimes grown after wheat. The amount of land occupied by durum wheat is not declared in any of the available documents. Certainly, Nigeria imports €38 million per year of durum wheat grain to be converted into pasta for the national and export market (€41 million worth). Hence, the local industry could certainly benefit from an increase in national production. Considering that the area cultivated with rice exceeds 2.9 million ha and that irrigation water is readily available in many parts of the countries, it certainly suggests great potential for expansion.

Similarly, Guinea is a large importer of durum grain (for €17 million EUR) but none is currently produced on the 1.6 million ha of rice cultivation. Côte d'Ivoire is the largest exporter of pasta (€32 million worth per year) but also one of the largest importers of durum grain (€28 million worth), with no production of wheat recorded on the 790,000 ha of rice cultivation (Table 1).

In summary, West African countries have the potential to convert their off-season of their 7.2 million ha of rice fields into durum wheat cultivation, instead of having an unproductive winter fallow. New, super-early and heat tolerant varieties have been developed, tested and confirmed along the Senegal River [87] and their seed is readily available through the CGIAR WHEAT program. Their cultivation could turn an annual import market of €185 million worth of grain and almost €200 million worth of pasta into a national income to improve industrialization, create jobs and reduce poverty in rural areas.

6. Southern and Central Africa Durum Wheat Use in the Industry with Limited Cultivation

Southern and Central African countries cultivate 1.6 million ha of rice and 0.65 million ha of wheat. Unfortunately, data on wheat cultivation in Central Africa are few and unsubstantial. Among Southern African countries, durum wheat is cultivated on just 26,500 ha, mostly in South Africa and Zimbabwe. The most widely cultivated varieties are the 'Desert' durum developed in Arizona and California, with "Kronos" (Arizona Plant Breeders male sterile-facilitated recurrent selection population selection) as the preferred one (Table 2). All countries obtain yields above 4 t ha^{-1}, which only partially meets the national industry demand. Still, part of the grains is exported for generating an income of €38 million.

All countries combined imported €160 million worth of durum grain in 2013 (Table 1). The largest importers of grain were Malawi, South Africa and Zimbabwe, which use it to sustain their national pasta industry. In fact, South Africa utilizes the grain to generate pasta for re-export with a value addition of over €11 million, while Cameroon reaches €1.4 million of pasta exports annually. Interestingly, some SSA countries do not apply import taxes on durum wheat, which in turn has promoted cases of illegal false labelling of bread wheat grain as durum wheat to avoid custom costs [25]. The import of pasta products in 2013 was €108 million worth and the biggest importers were South Africa and Madagascar, with €31 and 26 million worth, respectively. Therefore, a business opportunity exists for the local pasta industry, while creating the chance for growers to improve their livelihoods. Considering an average price per ton of durum wheat grain of €300 on local markets and attainable yields of 3 ton ha^{-1}, approximately 160,000 ha of the currently cultivated 650,000 ha of bread wheat would need to be converted to fill the production gap. Obviously, the reduction of bread wheat would in turn open a gap in the availability of national bread flour, pushing the country to further imports. However, import prices of bread wheat flour is significantly cheaper than durum wheat imports, especially when considering that durum wheat production is a trade that does not require government subsidies to be profitable. Hence, the national economy would overall benefit from a production shift toward durum wheat, as long as this does not upset the higher price paid for semolina. Furthermore, durum-bread wheat flour blends are commonly used in North Africa for the baking of affordable and protein-rich breads.

A second consideration is in regard to the spread of diseases. In fact, South Africa has been monitoring a growing threat of Karnal Bunt disease [94], while Uganda is the first country where the devastating stem rust race Ug99 was observed, before it spread to the neighbouring countries [95]. Both of these diseases affect prevalently bread wheat, while durum wheat has thus far remained resistant [96,97]. Hence, replacement of bread wheat by durum wheat would not only have a potential valuable effect on the economy but also reduce the incidence of damaging diseases on the wheat crop. Alternatively, durum wheat could be cultivated on part of the 1.8 million ha dedicated to rice during the fallow off-season period, assuming that adequate rainfall or irrigation water is available. This could be the case for Madagascar, where durum wheat could be cultivated during the off-season in the same terrace fields grown with paddy rice [98]. In fact, a recent study on wheat suitability in SSA [23] using geospatial analysis revealed that Angola, Mozambique, Zambia and Zimbabwe are the countries with the largest potential extension of suitable land for establishing wheat production. The suitable mega environments identified were highlands with high rainfall and frequent diseases (ME2A [99]) and drought prone rainfall with cold winter months (ME4A).

7. Durum Wheat Cultivation in the Saharan Oases: A Staple Food of Tradition

The Sahara oases are unique environments that remained impervious to modernization. In this review, both types of oases are considered; those areas of desert where water surfaces from the soil or where it can be collected by human activities through dams (*barrage*) or other methods as defined by Zaharieva et al. [100]. Semi-nomadic tribes live in these locations and developed self-sustaining agricultural systems based on the sporadic rainfalls and underground or aboveground water accumulations. Several major oases can be found in SSA in Chad, Mali, Mauritania, Niger and Sudan but also in Algeria, Egypt, Libya, Morocco and Tunisia. There are no extensive records of the total area cultivated. The Saharan oases are estimated at a total surface of 900,000 ha, of which approximately half is used for intensive agriculture [100]. Further, average sizes for oases are between 5 and 200 ha of cultivated land, depending on the abundance of yearly rainfall or available percolated water and can sustain the life of up to 1000 people per oasis [101]. In Mauritania, 350 oases account for a total surface of cultivation by wheat (bread and durum) of over 2000 ha [102]. Roughly the same area is cultivated in the oases of Mali [103], while the five largest oases in Algeria (Ghardaia region) account for 2200 ha of cereal culture [104] and in Morocco the major oasis region of Errachidia cultivates an approximate 5000 ha of cereals [105]. Cultivated crops include sorghum and millet as rainfed crops, both types of wheat and cowpea as irrigated crops. Larger oases have access to a constant water supply, allowing irrigation by pivot or drip irrigation, such as in the regions of east of Morocco and Algeria [106]. In these cases, water is pumped as needed and wheat is often cultivated among date palms with the moisture used by both cultures. In most other cases, large quantities of water are available only during specific times of the year and to collect it in sufficient amounts for cultivation, it is necessary to build temporary dams with clay, sand and stones. The dam is then opened at the beginning of the winter and as the water recedes, holes are dug into the mud and cereal grains are placed inside (Figure 3). Growing on residual moisture and with high temperatures, the yields rarely exceed 0.5 t ha^{-1}, while under pumped irrigation yields of 4 to 5 t ha^{-1} are common [107].

Figure 3. Wheat cultivation in oasis in Mauritania. (**a**) Holes in the mud for the planting of durum wheat as the water retreats; (**b**) Gradient on plant maturity caused by the difference in planting time following the retreat of the water.

The farmers of the desert cultivate mostly wheat biotypes of unique morphology defined as *oasiensis* types, which represent mixtures of several tetraploid and hexaploid wheat species (for review see Reference [108]). Durum wheat cultivation in the oases dates to the initial trade routes between the Nile Valley and West Africa [100]. Several traditional dishes are made from this crop and its straw is very important as feed for the small ruminants and camels. The 'Alkama Binka' is one of the most frequently found landraces in the Saharan oases of Algeria and Morocco [109]. Modern cultivars have also been introduced, such as 'Waha' (syn. 'Cham1', Plc/Ruff//Gta/Rtte) in Algeria and 'Karim' in Mauritania and their superior yields are causing a contraction in the use of landraces (Table 2). The wealth of genetic diversity of germplasm from the Saharan oases has been recognized by several authors and several calls for better collection and conservation have been made but with limited success [108]. In consideration of the harsh environment where these landraces thrive and the fact that durum production will be increasingly stressed due to climate issues [110], they certainly represent a valuable resource of useful alleles for heat, drought and salinity tolerance, which can be deployed in breeding for stress adaptation. Furthermore, the oases represent fragile ecosystems, where land availability is dependent on rainfall and maximum yields per unit of land are more critical than anywhere else. In that sense, the introduction of modern agronomy and irrigation practices, in integration with targeted breeding efforts could deliver true game changers. Alternatively, the reduced available land surface could be used as an advantage to generate very exclusive durum products. In fact, the 'rarity' could be exploited through well integrated value chains to deliver products at elevated prices on the occidental markets, as is already the case for the oases dates. Considering that oases produce less than 5% of their needs in cereals [104] and the rest is purchased from neighbouring towns, the possibility of generating larger incomes would be a suitable strategy to tackle famine.

In that sense, the already high value of durum grains could be further exploited via smart-marketing to increase the revenues.

8. Future Prospects: A South-South Collaboration to Expand Durum Wheat Cultivation in Africa

All of Africa accounts for an annual import of €4.1 billion worth of durum grain to supply the national pasta and couscous market. These are mostly imported to North Africa (NA) from Canada, USA and Turkey (Table 1). North Africa already cultivates durum wheat on 2.9 million ha and the area for further expansion is limited. This opens an opportunity for SSA to gain access to an €3.7 billion annual market by filling part of the grain needs of NA. The current area dedicated to wheat cultivation in SSA is limited to 2.6 million ha, mostly in Ethiopia, South Africa and Sudan. In Ethiopia, new interest has sprung toward the promotion of industrial crops such as durum wheat to provide the local manufacturers with prime raw material without the need of relying on expensive imports. In addition, urbanization has shifted the food habits of many countries and pasta has gained steadily in appreciation by African consumers. Furthermore, the case presented for cultivation of durum wheat in rotation with rice along the Senegal River, matches what is already customary on over 10 million ha of wheat-rice or wheat-rice-rice rotations in India [90]. In that sense, there is large potential for wheat expansion on the 9.1 million ha of rice land in SSA. Since further expansion of the wheat areas will require additional investments and will face the risk of reduced yields, it appears logical to seek the wheat type that would provide the maximum monetary return for unit of land converted. Durum wheat in this case would represent an ideal cash crop to help reduce poverty in SSA. For comparison, the average import prices of major cereals to South Africa [111] for the year 2015 were at: US$ 502 t^{-1} aromatic rice, US$ 330 t^{-1} durum wheat, US$ 278 t^{-1} malt barley, US$ 209 t^{-1} hard red bread wheat, US$ 171 t^{-1} sorghum and US$ 150 t^{-1} feed maize. While it is true that import prices change for each country based on access to trade, existence of infrastructure and specific import policies, South Africa provides a good example of a reactive trading nation in SSA. On this basis, it is evident that durum wheat remains one of the most income advantageous winter cereals, significantly more expensive than bread wheat and malt barley. However, to succeed in the utilization of the financial return of this crop, it is necessary to have a well-integrated value chain capable of delivering profitable economic returns to farmers. The example of the value chain in the Oromia region of Ethiopia could be repeated in several other regions and should provide a good guideline to follow for out-scaling to other countries. Still, the industrial machinery and the strategy for production need to be harmonized among African countries to generate a fair and vibrant market. The desire for semolina-based food is expected to increase in the years to come [112] but the national industry will be successful in targeting the demand only if their products can compete not just in price but also in quality with the imported ones. In that sense, great traditional and modern knowledge for cultivation and production of this crop exist already in North Africa and Ethiopia. Breeding programs for this crop have been successful in targeting the harsh drought conditions of North Africa and the disease pressure in Ethiopia. In order to expand the production of this crop to non-traditional territories, the expertise gathered there could be transferred to SSA in the form of novel and adapted varieties. It is therefore desirable that Ethiopian breeders could produce varieties well adapted to the SSA mega-environment of type 2A, with high rainfall and high disease pressure. Instead, Egyptian breeders could help in delivering varieties targeted to the hot and irrigated areas of mega-environment type ME1, such as West Africa and Sudan. The other North African countries could target ME4A, with low rainfall and cold winters, as well as help in the further development of the Saharan oases. Altogether, this envisioned South-South collaboration could ensure that varieties developed in traditional durum growing areas such as North Africa and Ethiopia, would adapt to the conditions of the southern partners. Harvests could then be sold to those African countries with strong pasta industries and the finished semolina products would be sold all over Africa. This integrated value chain would ensure a steep increase in monetary circulation and an overall reduction in the poverty of Africa. Recent publicly funded projects like Africa Rising [113],

SARD-SC [114], TAAT [115] and U-Forsk2018 have targeted the increase in production of wheat in SSA and created the basis to hope for a comprehensive "durum wheat revolution" in SSA.

Author Contributions: Writing—original draft preparation, A.T.S., F.M.B., T.C., W.L.; Writing—review and editing, F.M.B., A.T.S., M.v.G., and K.S.; supervision, F.M.B., M.v.G., R.O.; project administration, R.O.; funding acquisition, F.M.B., R.O.

Funding: This research was funded by the Swedish Research Council (Vetenskapsrådet) U-forsk 2013-6500 "Deployment of molecular durum breeding to the Senegal basin: capacity building to face global warming" and U-forsk 2017-05522 "Genomic prediction for breeding durum wheat along the Senegal River Basin".

Conflicts of Interest: The authors declare no conflict of interest. The founding sponsors had no role in the writing of the manuscript and in the decision to publish the results.

References

1.   Chris, G. World Durum Outlook. Available online: http://www.internationalpasta.org/resources/IPO%20BOARD%202013/2%20Chris%20Gillen.pdf (accessed on 7 April 2017).

2.   Statistic Canada. Canada: Outlook for Principal Field Crops. Available online: http://www.agr.gc.ca/eng/industry-markets-and-trade/statistics-and-market-information/by-product-sector/crops-industry/outlook-for-principal-field-crops-in-canada/canada-outlook-for-principal-field-crops-february-16-2016/?id=1455720699951 (accessed on 7 April 2017).

3.   USDA Foreign Agricultural Service. *Grain and Feed Annual*; GAIN Report No: TR5016; USDA Foreign Agricultural Service: Ankara, Turkey, 2015.

4.   Nagarajan, S. Quality characteristics of Indian wheat. In *Future of Flour*; Popper, L., Schäfer, W., Freund, W., Eds.; AgriMedia GmbH: Clenze, Germany, 2006; pp. 79–86.

5.   Le Lamer, O.; Rousselin, X. The durum wheat market. In *Studies of FranceAgriMer*; Bova, F., Ed.; FranceAgriMer: Montreuil-sous-Bois CEDEX, France, 2011; pp. 1–46.

6.   Bonjean, A.P.; Angus, W.J.; van Ginkel, M. *The World Wheat Book: A History of Wheat Breeding*; Lavoisier: Paris, France, 2016; Volume 3.

7.   Al-Issa, T.A.; Samarah, N.H. Tillage practices in wheat production under rainfed conditions in Jordan: An economic comparison. *World J. Agric. Sci.* **2006**, *2*, 322–325.

8.   Karam, F.; Kabalan, R.; Breidi, J.; Rouphael, Y.; Oweis, T. Yield and water-production functions of two durum wheat cultivars grown under different irrigation and nitrogen regimes. *Agric. Water Manag.* **2009**, *96*, 603–615. [CrossRef]

9.   El-Areed, S.; Nachit, M.M.; Hagaras, A.; El-Sherif, S.; Hamouda, M. Durum wheat breeding for high yield potential in Egypt. In *Proceedings of the International Symposium on Genetics and Breeding of Durum Wheat*; Porceddu, E., Damania, A.B., Qualset, C.O., Eds.; CIHEAM: Bari, Italy, 2014; pp. 291–294.

10. Juarez, B.; Wolf, D. Grain and feed annual Mexico. In *Global Agricultural Information*; USDA Foreign Agricultural Service: Washington, DC, USA, 2015; pp. 1–17.

11. John, K. Durum Wheat Production. Available online: http://www.nvtonline.com.au/wp-content/uploads/2013/03/Crop-Guide-NSW-Durum-Wheat-Production.pdf (accessed on 24 December 2016).

12. Evan School Policy Analysis and Research (EPAR). Wheat Value Chain: Ethiopia. Available online: https://evans.uw.edu/sites/default/files/EPAR_UW_204_Wheat_Ethiopia_07272012.pdf (accessed on 24 December 2016).

13. Elias, E.M. Durum wheat products. In *Durum Wheat Quality in the Mediterranean Region*; Di Fonzo, N., Kaan, F., Nachit, M., Eds.; CIHEAM: Zaragoza, Spain, 1995; Volume 22, pp. 23–31.

14. International Pasta Organisation (IPO). The Truth about Pasta Toolkit. Available online: http://www.internationalpasta.org/index.aspx?id=47 (accessed on 27 December 2016).

15. Simoes, A.J.G.; Hidalgo, C.A. The Economic Complexity Observatory: An analytical tool for understanding the dynamics of economic development. In *Scalable Integration of Analytics and Visualization*; San Francisco, CA, USA, 2011.

16. Food Price Index (FPI). Available online: http://www.fao.org/worldfoodsituation/foodpricesindex/en/ (accessed on 22 December 2016).

17. Food and Agriculture Organization (FAO). *FAO Statistical Yearbook 2014: Africa*; Tijani, B., Gennari, P., Eds.; FAO: Rome, Italy, 2014.

18. Food and Agriculture Organization (FAO). Eritrea: Country Report. In Proceedings of the FAO International Technical Conference on Plant Genetics, Leipzig, Germany, 17–23 June 1996.

19. Bill and Melinda Gates Foundation (BMGF). Multi Crop Value Chain Phase II: Wheat Ethiopia. Available online: https://agriknowledge.org/files/8k71nh141 (accessed on 10 January 2017).

20. What Farming in Kenya. Available online: https://informationcradle.com/kenya/wheat-in-kenya/ (accessed on 15 May 2019).

21. Saunders, D.A.; Hettel, G.P. *Wheat in Heat-Stressed Environments: Irrigated, Dry Areas and Rice-Wheat Farming Systems*; UNDP; ARC; BARI; CIMMYT: Mexico, 1993; ISBN 968-6127-87-9.

22. Mistretta, G. Angola: Alcune caratteristiche del mercato della pasta. Ricadute sul Prodotto Italiano. Ambassador letter on 06 November 2013 to InfoMercatiEsteri.it. Available online: http://www.infomercatiesteri.it/public/images/paesi/4/files/PASTA%20di%20grano.doc (accessed on 15 May 2019).

23. Negassa, A.; Shiferaw, B.; Koo, J.; Sonder, K.; Smale, M.; Braun, H.J.; Gbegbelegbe, S.; Guo, Z.; Hodson, D.; Wood, S.; et al. *The Potential for Wheat Production in Africa: Analysis of Biophysical Suitability and Economic Profitability*; CIMMYT: DF, Mexico, 2013.

24. Ssekitoleko, G.W.; Vogel, W.O. Wheat marketing and pricing policy in Uganda: A critical assessment. In *Proceedings of Theninth Regional Wheat Workshop for Eastern, Central and Southern Africa, Addis Abeba, Ethiopia, 2–6 October 1995*; Tanner, D.G., Payne, T.S., Abdalla, O.S., Eds.; CIMMYT: Mexico City, Mexico, 1995.

25. United States Agency for International Development (USAID). *Staple Foods Value Chain Analysis: Country Report Zambia*; Chemonics International Inc.: Washington, DC, USA, 2009.

26. Kayoed-Anglade, S.; Aaron, R. Projet du Moulin Modern du Mali. Available online: https://www.afdb.org/fileadmin/uploads/afdb/Documents/Environmental-and-Social-Assessments/Mali_-_Projet_du_moulin_moderne_du_Mali_-_R%c3%a9sum%c3%a9_PGES.pdf (accessed on 4 January 2017).

27. Factfish: Research Made Simple. Available online: www.factfish.com (accessed on 15 May 2019).

28. Index Mundi. Available online: www.indexmundi.com (accessed on 15 May 2019).

29. Hakan, O.; Willcox, G.; Graner, A.; Salamini, F.; Kilian, B. Geographic distribution and domestication of wild emmer wheat (*Triticum dicoccoides*). *Genet. Resour. Crop Evol.* **2010**, *58*, 11–53.

30. Benbelkacem, A.; Khaldounl, A.; Djenadi, C. The history of wheat breeding in Algeria. In *World Wheat Book: A History of Wheat Breeding*; Bonjean, A.P., Angus, W.J., van Ginkel, M., Eds.; Lavoisier: Paris, France, 2016; Volume 3, pp. 477–495.

31. National Research Council. *Lost Crops of Africa: Volume I: Grains*; The National Academies Press: Washington, DC, USA, 1996.

32. Luo, M.C.; Yang, Z.L.; You, F.M. The structure of wild and domesticated emmer wheat populations, gene flow between them and the site of emmer domestication. *Theor. Appl. Genet.* **2007**, *114*, 947–959. [CrossRef]

33. Kabbaj, H.; Sall, A.T.; Al-Abdallat, A.; Geleta, M.; Amri, A.; Filali-Maltouf, A.; Belkadi, B.; Ortiz, R.; Bassi, F.M. Genetic Diversity within a Global Panel of Durum Wheat (*Triticum durum*) Landraces and Modern Germplasm Reveals the History of Alleles Exchange. *Front. Plant Sci.* **2017**, *8*, 1277. [CrossRef] [PubMed]

34. Mengistu, D.K.; Kirosa, Y.A.Y.; Pè, M.E. Phenotypic diversity in Ethiopian durum wheat (*Triticum turgidum* var. durum) landraces. *Crop J.* **2015**, *3*, 190–199. [CrossRef]

35. Vavilov, N.I. The origin, variation, immunity and breeding of cultivated plants, trans from the Russian by Starr Chester, K., Waltham, Mass.: Chronica Botanica. *Science* **1951**, *115*, 364.

36. Pecetti, L.; Annicchiarico, P.; Damania, A.B. Biodiversity in a germplasm collection of durum wheat. *Euphytica* **1992**, *60*, 229–238.

37. Ficco, D.B.M.; Mastrangelo, A.M.; Trono, D.; Borrelli, G.M.; De Vita, P.; Fares, C.; Beleggia, R.; Platani, C.; Papa, R. The colors of durum wheat: A review. *Crop Pasture Sci.* **2014**, *65*, 1–15. [CrossRef]

38. Mass, F.B.; Paterson, F.L.; Foster, J.E.; Ohm, H.W. Expression and inheritance of resistance of ELS 6404-160 durum wheat to Hessian Fly. *Crop Sci.* **1989**, *29*, 23–28. [CrossRef]

39. Belay, G.; Merker, A.; Tesfaye, T. Cytogenetic studies in Ethiopian landraces of tetraploid wheat (Tritium turgidum L.), spike morphology and ploidy level and karyomorphology. *Hereditas* **1994**, *121*, 45–52. [CrossRef]

40. Bekele, E. Analysis of regional patterns of phenotypic diversity in the Ethiopian tetraploid and hexaploid wheats. *Hereditas* **1984**, *100*, 131–154. [CrossRef]

41. Negassa, M. Estimates of phenotypic diversity and breeding potential of Ethiopian wheats. *Hereditas* **1986**, *104*, 41–48. [CrossRef]

42. Tesemma, T.; Belay, G. Aspects of Ethiopian tetraploid wheats with emphasis on durum wheat genetics and breeding research. In *Wheat Research in Ethiopia: A Historical Perspective*; Hailu, G., Tanner, D.G., Mengistu, H., Eds.; CIMMYT: El Batan, Mexico, 1991; pp. 47–72.

43. Belay, G.; Tesfaye, T.; Becker, H.C.; Merker, A. Variation and interrelationships of agronomic traits in Ethiopian tetraploid wheat landraces. *Euphytica* **1993**, *71*, 181–188. [CrossRef]

44. Bechere, E.; Belay, G.; Mitiku, D.; Merker, A. Phenotypic diversity of tetraploid wheat landraces from northern and north-central regions of Ethiopia. *Hereditas* **1996**, *124*, 165–172. [CrossRef]

45. Tsegaye, S.; Tesemma, T.; Belay, G. Relationships among tetraploid wheat (*Triticum turgidum* L.) landrace populations revealed by isozyme markers and agronomic traits. *Theor. Appl. Genet.* **1996**, *93*, 600–605. [CrossRef]

46. Eticha, F.; Bekele, E.; Belay, G.; Börner, A. Phenotypic diversity in tetraploid wheats collected from Bale and Wello regions of Ethiopia. *Plant Genet. Resour.* **2005**, *3*, 35–43. [CrossRef]

47. Mengistu, D.K.; Kidane, Y.G.; Catellani, M.; Frascaroli, E.; Fadda, C.; Pe, M.E.; Dell'Acqua, M. High-density molecular characterization and association mapping in Ethiopian durum wheat landraces reveals high diversity and potential for wheat breeding. *Plant Biotechnol. J.* **2016**, *14*, 1800–1812. [CrossRef] [PubMed]

48. Mengistu, D.K.; Kidane, Y.G.; Fadda, C.; Pè, M.E. Genetic diversity in Ethiopian Durum Wheat (*Triticum turgidum* var durum) inferred from phenotypic variations. *Plant Genet. Resour.* **2016**, *16*, 39–49. [CrossRef]

49. Tsegaye, B.; Berg, T. Utilization of durum wheat landraces in East Shewa, central Ethiopia: Are home uses an incentive for on-farm conservation? *Agric. Hum. Values* **2007**, *24*, 219–230. [CrossRef]

50. Fassil, K.; Sehaye, Y.T.; McNeilly, T. Diversity of durum wheat (Triticum durum Desf.) at in situ conservation sites in North Shewa and Bale, Ethiopia. *J. Agric. Sci.* **2001**, *136*, 383–392.

51. Israel, O.K.; Bhandari, B.; Guilherme de Azevedo, S.; Natarajan, K.; Gezu, G.; Gashu, S.; Desalegn, T. *Global Study on CBM and Empowerment, Ethiopia Exchange Report*; Wageningen University: Wageningen, The Netherlands, 2010; p. 42.

52. Tesemma, T.; Belay, G.; Worede, M. Morphological diversity in tetraploid wheat landrace populations from the central highlands of Ethiopia. *Hereditas* **1991**, *114*, 171–176.

53. Tolera, A.; Tsegaye, B.; Berg, T. Effects of variety, cropping year, location and fertilizer application on nutritive value of durum wheat straw. *J. Anim. Physiol. Anim. Nutr.* **2007**, *92*, 121–130. [CrossRef]

54. Bezabeh, E.; Haregewoin, T.; Giorgis, D.H.; Daniel, F.; Belay, B. Change and growth rate analysis in area, yield and production of wheat in Ethiopia. *Res. J. Agric. Environ. Manag.* **2015**, *4*, 189–191.

55. Bassi, F.M.; Chiari, T.; Verma, R.P.S.; Kumar, S. Crop Diversification: The 'Old' Weapon in the Hands of Ethiopian Farmers, Borlaug Global Rust Initiative. Available online: http://www.globalrust.org/blog/crop-diversification-%E2%80%98old%E2%80%99-weapon-hands-ethiopian-farmers (accessed on 10 January 2017).

56. Global Milling. Ethiopia Battles Wheat Rust Disease Outbreak in Critical Wheat-Growing Regions. Available online: http://globalmilling.com/ethiopia-battles-wheat-rust-disease-outbreak-in-critical-wheat-growing-regions/ (accessed on 27 December 2016).

57. Tesemma, T.; Mohammed, J. Review of wheat breeding in Ethiopia. *Ethiopian J. Agric. Sci.* **1982**, *1*, 11–24.

58. Nastasi, V. *Wheat Production in Ethiopia*; Information Bulletin on the Near East Wheat and Barley Improvement and Production Project; FAO: Rome, Italy, 1964; Volume 1, pp. 13–24.

59. Tesemma, T. Durum wheat breeding in Ethiopia. In *Fifth Regional Wheat Workshop for Eastern, Central and Southern Africa and the Indian Ocean*; van Ginkel, M., Tanner, D.G., Eds.; CIMMYT: El Batan, Mexico, 1988; pp. 18–22.

60. Consultative Group on International Agricultural Research (CGIAR). Virulent new strains of Ug99 stem rust, a deadly wheat pathogen. *Science Daily*, 28 May 2010.

61. Ministry of Agriculture (MoA). Plant Variety Release, Protection and Seed Quality Control Directorate. *Crop Var. Regist.* **2013**, *16*, 301.

62. Leta, G.; Belay, G.; Walelign, W. Nitrogen fertilization effects on grain quality of durum wheat (*Triticum turgidum L.* var. durum) varieties in central Ethiopia. *Agric. Sci.* **2013**, *4*, 123–130.

63. Mekuria, T.; Wasihun, L.; Shitaye, H.; Asenafi, G. Durum wheat (Triticum durum Desf) Variety "Utuba" Performance in Ethiopia. *Agric. Res. Technol. Open Access J.* **2018**, *18*, 556063. [CrossRef]

64. De Groote, H.; Gitonga, Z.; Mugo, S.; Walker, T.S. Assessing the effectiveness of maize and wheat improvement from the perspectives of varietal output adoption in east and southern Africa. In *Crop Improvement, Adoption and Impact of Improved Varieties in Food Crops in Sub Saharan Africa*; Walker, T.S., Alwang, J., Eds.; CABI: Egham, UK, 2015; pp. 207–227.

65. Dawit, A. *Farmer-Based Seed Multiplication in the Ethiopian Seed System: Approaches, Priorities and Performance*; FAC Working Paper 36; Future Agricultures Consortium: Brighton, UK, 2011.

66. Bishaw, Z. Wheat and Barley Seed Systems in Ethiopia and Syria. Ph.D. Thesis, Wageningen University, Wageningen, The Netherlands, 2004; pp. 1–401.

67. Joshi, A.K.; Azab, M.; Mosaad, M.; Braun, H.J. Delivering rust resistant wheat to farmers: A step towards increased food security. *Euphytica* **2011**, *179*, 187–196. [CrossRef]

68. Ojiewo, C.O.; Kugbei, S.; Bishaw, Z.; Rubyogo, J.C. Community seed production. FAO & ICRISAT: 2015. In *Workshop Proceedings, 9–11 December 2013*; FAO: Rome, Italy; ICRISAT: Addis Ababa, Ethiopia, 2013; 176p.

69. Bishaw, Z.; Niane, A.A. Are farmer-based seed enterprises profitable and sustainable? Experiences of VBSEs from Afghanistan. In Proceedings of the Community Seed Production, Rome, Italy, 9–11 December 2013; FAO: Rome, Italy; ICRISAT: Addis Ababa, Ethiopia, 2013; pp. 55–64.

70. Board on Science and Technology for International Development (BOSTID). *Lost Crops of Africa, Vol. I Grains*; National Academic Press: Washington, DC, USA, 1996.

71. Central Statistical Agency (CSA). *Agricultural Sample Survey 2013/2014, Statistical Bulletin No 536*; Federal Democratic Republic of Ethiopia Central Statistical Agency: Addis Ababa, Ethiopia, 2014; pp. 1–121.

72. D'Andrea, C.A.; Haile, M. Traditional emmer processing in highland Ethiopia. *J. Ethnobiol.* **2002**, *22*, 179–217.

73. Cooper, R. Re-discovering ancient wheat varieties as functional foods. *J. Tradit. Complement. Med.* **2015**, *5*, 138–143. [CrossRef] [PubMed]

74. Stallknecht, G.F.; Gilbertson, K.M.; Romey, J.E. Alternative wheat cereals as good grains: Einkorn, emmer, spelt, kamut and triticale. In *Progress in New Crops*; Janick, J., Ed.; ASHS Press: Alexandria, VA, USA, 1997; pp. 156–170.

75. Negassa, M. Possible new genes for resistance to powdery mildew, septoria glume blotch and leaf rust of wheat. *Plant Breed.* **1987**, *98*, 37–46. [CrossRef]

76. Damania, A.B.; Hakim, S.; Moualla, M.Y. Evaluation of variation in *T. dicoccum* for wheat improvement in stress environment. *Hereditas* **1992**, *116*, 163–166. [CrossRef]

77. Robinson, J.; Skovmand, B. Evaluation of emmer wheat and other Triticeae for resistance to Russian wheat aphid. *Genet. Resour. Crop Evol.* **1992**, *39*, 159–163.

78. Naod, B.; Fininsa, C.; Badebo, A. Sources of stem rust resistance in Ethiopian tetraploid wheat accessions. *Afr. Crop Sci. J.* **2007**, *15*, 51–57.

79. Oliver, R.E.; Cai, X.; Friesen, T.L.; Halley, S.; Stack, R.W.; Xu, S.S. Evaluation of Fusarium head blight resistance in tetraploid wheat (*Triticum turgidum* L.). *Crop Sci.* **2008**, *48*, 213–222. [CrossRef]

80. Ethiopian Revenues and Customs Authority (ERCA). 2014. Available online: http://www.erca.gov.et/index.php (accessed on 5 January 2017).

81. Geleto, T.; Tanner, D.G.; Mamo, T.; Gebeyehu, G. Response of rainfed bread and durum wheat to source, level and timing of nitrogen fertilizer on two Ethiopian vertisols: N uptake, recovery and efficiency. *Fertil. Res.* **1996**, *44*, 195–204. [CrossRef]

82. Biggeri, M.; Ciani, F.; Ferrannini, A. Aid effectiveness and multilevel governance: The case of a value chain development project in rural Ethiopia. *Eur. J. Dev. Res.* **2016**, *29*, 843–865. [CrossRef]

83. Ministry of Agriculture (MOA). Plant Variety Release, Protection and Seed Quality Control Directorate. *Crop Var. Regist.* **2015**, *17*, 330.

84. Biggeri, M.; Burchi, F.; Ciani, F.; Herrmann, R. Linking small-scale farmers to the durum value chain in Ethiopia: Assessing the effects on production and wellbeing. *Food Policy* **2018**, *79*, 77–91. [CrossRef]

85. Falaki, A.M.; Mohammed, I.B. Performance of some durum wheat varieties at Kadawa, Kano state of Nigeria. *Bayero J. Pure Appl. Sci.* **2011**, *4*, 48–51. [CrossRef]

86. Agriculture in Mauritania. Encyclopedia of the Nations—Information about Countries of the World, United Nations and World Leaders. Available online: http://www.nationsencyclopedia.com/Africa/Mauritania-AGRICULTURE.html (accessed on 15 December 2016).

87. Sall, A.T.; Kabbaj, H.; Cisse, M.; Gueye, H.; Ndoye, I.; Maltouf, A.F.; El-Mourid, M.; Ortiz, R.; Bassi, F.M. Heat Tolerance of Durum Wheat (*Tritcum durum* Desf.) Elite Germplasm Tested along the Senegal River. *J. Agric. Sci.* **2018**, *10*, 217–233. [CrossRef]

88. Food and Agriculture Organization (FAO). *Senegal: Irrigation Market Brief*; Report No. 26; FAO: Rome, Italy, January 2016.

89. Manikowski, S.; Strapasson, A. Sustainability Assessment of Large Irrigation Dams in Senegal: A Cost-Benefit Analysis for the Senegal River Valley. *Front. Environ. Sci.* **2016**, *4*, 18. [CrossRef]

90. Mahajan, A.; Gupta, R.D. *Integrated Nutrient Management (INM) in a Sustainable Rice-Wheat Cropping System*; Springer: Heidelberg, Germany, 2009; p. 268.

91. Club du Sahel (CILSS). *Development of Irrigated Agriculture in Upper Volta: Proposals for a Second Programme 1980-85*; Permanent Interstate Committee for Drought Control in the Sahel: Niamey, Republic of Niger, 1979.

92. Chabrolin, R. Programme pour les Recherches Concernant le blé au Mali; Fonds IRD: B22431. 1967.

93. USDA Foreign Agricultural Service. 2018 Grain and Feed Annual Nigeria-USDA GAIN Reports. Available online: https://gain.fas.usda.gov/Recent%20GAIN%20Publications/Grain%20and%20Feed%20Annual_Lagos_Nigeria_4-12-2018.pdf (accessed on 5 March 2019).

94. Stansbury, C.D.; Pretorius, Z.A. Modelling the potential distribution of Karnal bunt of wheat in South Africa. *S. Afr. J. Plant Soil* **2001**, *18*, 159–168. [CrossRef]

95. Singh, R.P.; Hodson, D.P.; Huerta-Espino, J.; Jin, Y.; Bhavani, S.; Njau, P.; Herrera-Foessel, S.; Singh, P.K.; Singh, S.; Govindan, V. The emergence of Ug99 races of the stem rust fungus is a threat to World wheat production. *Annu. Rev. Phytopathol.* **2011**, *49*, 465–481. [CrossRef]

96. Duveiller, E.; Mezzalama, M. *Karnal Bunt: Screening for Resistance and Distributing KB Free Seed*; CIMMYT: El Batan, Mexico, 2009.

97. Letta, T.; Olivera, P.; Maccaferri, M.; Jin, Y.; Ammar, K.; Badebo, A.; Noli, E.; Crossa, J.; Tuberosa, R. Genome-wide search of stem rust resistance loci at the seedling stage in durum wheat. *Plant Genome* **2014**, *7*, 1–13. [CrossRef]

98. Ravelomamtsoa, S.H.; Randrianaivoarivony, J.M.; Ramalanjaona, V.L. Overview of wheat in Madagascar. In Proceedings of the Wheat for Food Security in Africa, Addis-Ababa, Ethiopia, 8–12 October 2012.

99. Rajaram, S.; van Ginkel, M.; Fischer, R.A. CIMMYT's wheat breeding mega-environments (ME). In Proceedings of the 8th International Wheat Genetics Symposium, Beijing, China, 19–24 July 1994.

100. Zaharieva, M.; Bonjean, A.; Monneveux, P. Alert: Saharan Oases wheat genetic resources in danger. In *World Wheat Book: A History of Wheat Breeding*; Bonjean, A.P., Angus, W.J., van Ginkel, M., Eds.; Lavoisier: Paris, France, 2016; Volume 3, pp. 543–588.

101. Saharan Development. Agriculture and Farming. Available online: http://www.sahara-developpement.com/Western-Sahara/AgricultureEtElevage--117.aspx (accessed on 26 December 2016).

102. Food and Agriculture Organization (FAO). La Mauritanie et la FAO: Partenariat pour Améliorer la Résilience et Renforcer la Sécurité Alimentaire et Nutritionnelle. Available online: http://www.fao.org/3/a-au199f.pdf (accessed on 25 December 2016).

103. Food and Agriculture Organization (FAO)-ISESCO. In Proceedings of the International Workshop on Globally Important Agriculture Heritage Systems (GIAHS) for the Islamic Countries, Rome, Italy, 4–5 November 2014.

104. Houichiti, R.; Bissati, S.; Bouammar, B. Oasis agriculture and food insecurity in Algeria: The case of Ghardaia region. *Pensee J.* **2014**, *76*, 1–7.

105. Ait, H. Systemes de production et stratégies des agriculteurs dans les oasis de la region d'Errachidia au Maroc. *New Medit.* **2003**, *2*, 37–43.

106. El Abbass, S. Systèmes d'irrigation dans les oasis de Mauritanie: Problèmes de pompage et tentatives de réalimentation des nappes phréatiques. In Proceedings of the Journées Internationales sur l'Agriculture et la Gastronomie des Oasis, Elche, Spain, 14–15 October 2009.

107. Merouche, A.; Debaeke, P.; Messahel, M.; Kelkouli, M. Response of Durum wheat varieties to water in semi-arid Algeria. *Af. J. Agric. Res.* **2014**, *9*, 2880–2893.

108. Zaharieva, M.; Bonjean, A.; Monneveux, P. Saharan wheats: Before they disappear. Genet. *Resour. Crop Evol.* **2014**, *61*, 1065–1084. [CrossRef]

109. Benlaghlid, M.; Bouattoura, N.; Monneveux, P.; Borries, C. Etude de la diversité génétique et de la physiologie de l'adaptation au milieu. *Options Méditerranéennes* **1990**, *11*, 171–194.

110. Constantinidou, K.; Zittis, G.; Hadjinicolaou, P. Variations in the Simulation of Climate Change Impact Indices due to Different Land Surface Schemes over the Mediterranean, Middle East and Northern Africa. *Atmosphere* **2019**, *10*, 26. [CrossRef]
111. South African Revenue Service. Available online: www.sagis.org.za/sars.html (accessed on 7 February 2017).
112. William, A.R. Market Trends Category Analysis: A Look into the Pasta. Available online: www.preparedfoods.com/articles/103693-market-trends-category-analysis-a-look-into-the-pasta (accessed on 5 January 2017).
113. Karaimu, P. Africa Research in Sustainable Intensification for the Next Generation. Available online: https://africa-rising.net/ (accessed on 15 May 2019).
114. Hauser, J.F. Support to Agricultural Research for Development of Strategic Crops in Africa. Available online: http://sard-sc.org/ (accessed on 15 May 2019).
115. IITA. TAAT Launch Signifies New Day for African Agriculture. Available online: https://www.iita.org/news-item/taat-launch-signifies-new-day-african-agriculture/ (accessed on 15 May 2019).

Article

Loci Controlling Adaptation to Heat Stress Occurring at the Reproductive Stage in Durum Wheat

Khaoula El Hassouni [1,2], Bouchra Belkadi [2], Abdelkarim Filali-Maltouf [2], Amadou Tidiane-Sall [1,3], Ayed Al-Abdallat [4], Miloudi Nachit [1,2] and Filippo M. Bassi [1,*]

[1]   International Center for the Agricultural Research in Dry Area (ICARDA), Rabat 10000, Morocco
[2]   Faculty of sciences, University of Mohammed V, Rabat 10000, Morocco
[3]   Institut Sénégalais de Recherches Agricoles (ISRA), Saint-Louis 46024, Senegal
[4]   Faculty of Agriculture, The University of Jordan, Amman 11942, Jordan
*   Correspondence: f.bassi@cgiar.org; Tel.: +212614402717

Received: 27 June 2019; Accepted: 23 July 2019; Published: 30 July 2019

Abstract: Heat stress occurring during the reproductive stage of wheat has a detrimental effect on productivity. A durum wheat core set was exposed to simulated terminal heat stress by applying plastic tunnels at the time of flowering over two seasons. Mean grain yield was reduced by 54% compared to control conditions, and grain number was the most critical trait for tolerance to this stress. The combined use of tolerance indices and grain yield identified five top performing elite lines: Kunmiki, Berghouata1, Margherita2, IDON37-141, and Ourgh. The core set was also subjected to genome wide association study using 7652 polymorphic single nucleotide polymorphism (SNPs) markers. The most significant genomic regions were identified in association with spike fertility and tolerance indices on chromosomes 1A, 5B, and 6B. Haplotype analysis on a set of 208 elite lines confirmed that lines that carried the positive allele at all three quantitative trait loci (QTLs) had a yield advantage of 8% when field tested under daily temperatures above 31° C. Three of the QTLs were successfully validated into Kompetitive Allele Specific PCR (KASP) markers and explained >10% of the phenotypic variation for an independent elite germplasm set. These genomic regions can now be readily deployed via breeding to improve resilience to climate change and increase productivity in heat-stressed areas.

Keywords: heat stress; durum wheat; yield; tolerance; fertility; climate change; resilience

1. Introduction

Heat stress is a major environmental constraint to crop production. Terminal heat stress is defined as a rise in temperature that occurs between heading and maturity. When this stress matches with the reproductive phase of the wheat plant, it affects anthesis and grain filling, resulting in a severe reduction in yield [1]. High temperatures at the time of flowering cause floret sterility via pollen dehiscence [2], decrease photosynthetic capacity by drying the green tissues, and reduce starch biosynthesis [1,3]. These in turn result in a negative effect on grain number and weight [4–7]. The optimum growing temperature for wheat during pollination and grain filling phases is 21 °C [8,9], and for each increase of 1 °C above it is estimated a decline of 4.1% to 6.4% in yield [10]. Environmental temperatures have been increasing over the last century and more frequent heat waves are predicted in the next decades [11–13]. Therefore, breeding for tolerance to chronic as well as short term heat stress is a major objective worldwide [14–19]. Breeding selection would benefit by a better understanding of traits associated with tolerance to high temperatures, as well as the identification of the genomic regions controlling these traits.

In wheat, a large number of quantitative trait loci (QTLs) has been identified under heat stress via linkage analysis and genome-wide association study (GWAS) for yield, yield related traits, and some physiological traits such as chlorophyll content, chlorophyll fluorescence, and canopy temperature [20–27]. Grain number per spike and chlorophyll content were found to be the most critical traits for adaptation to warm conditions [24,25,28]. Heat stress reduces leaf chlorophyll content [29] affecting the amount of carbohydrates transported to the grains and final grain weight and size. High temperatures around anthesis reduce the number of grains per spike due to a decrease in spike growth and development, and an increase in ovules abortion [2,25,29,30]. To the best of our knowledge, molecular markers associated with heat tolerance are not generally used in wheat breeding programs [31–33]. The limited understanding of genes underlying physiological mechanisms and the regulation of yield components in wheat, and the lack of cloned major QTL for traits associated with heat tolerance has restricted the improvement in breeding for tolerance to this stress.

In the current study, a set of durum wheat lines were heat stressed by imposing a > 10 °C raise in maximum daily temperatures via the deployment of plastic tunnels at the time of flowering. GWAS studies allowed the identification of major QTLs controlling the adaptation to this stress and these were validated for marker assisted selection (MAS) in an independent germplasm set for rapid deployment via breeding.

2. Materials and Methods

2.1. Plant Material

A subset of 42 durum wheat inbred lines were selected from a global collection of 384 genotypes based on their similarity in flowering time and identified genetic diversity [34]. Briefly, the complete collection is highly diverse and includes 96 durum wheat landraces from 24 countries, and 288 modern lines from nine countries and two International research centers CIMMYT and ICARDA. The subset selected for this study includes 34 ICARDA and CIMMYT lines, five cultivars and one landrace. The list of the 42 genotypes and their details are provided in Table S1.

A second subset of 208 modern entries was also obtained from the global collection and field tested under severe high temperatures during 2014–2015 and 2015–2016 seasons along the Senegal River in Kaedi, Mauritania. Full details on this field experiment have been published in Sall et al. [35].

The third and final set was used for Kompetitive Allele Specific PCR (KASP) markers validation and it was composed of 94 ICARDA's elite lines that constituted the 2017 international nurseries 40th International Durum Yield Trial (IDYT) and 40th International Durum Observation Nurseries (IDON). This set was also tested at the station of Kaedi along the Senegal River in season 2015–2016.

2.2. Field Experiment Conditions and Phenotyping

The first subset of 42 entries was grown at Marchouch station (33°34′3.1″ N, 6°38′0.1″ W) in Morocco during two successive crop seasons (2015–2016 and 2016–2017). Each entry was sown in mid-November on a plot surface of 1.5 m^2 per genotype at a sowing density of 300 plants per m^2. The experiment was an alpha lattice with two replications, block size of six, and two treatments arranged in split-plot. Each six genotypes were arranged in close proximity to maximize competition between the genotypes, and compose one block of 9 m^2. Each block was surrounded by a border of barley to avoid border effect. Each block was spaced 1 m apart to allow the application of the plastic tunnel. The two treatments were normal rainfed conditions and plastic tunnel-mediated heat stress. The normal treatment followed standard agronomic practices with a base pre-sowing application of 50 Kg ha^{-1} of N, P, and K. At stage 15 of Zadok's (Z) scale herbicide was applied in a tank mixture (Pallas + Mustang at 0.5 L ha^{-1}) to provide protection against both monocots and dicots. At Z17 ammonium nitrate was provided to add 36 kg ha^{-1} of N and a final application of urea was used to add 44 kg ha^{-1} of N before booting (Z39). Weeds were also controlled mechanically to ensure clean plots. The soil of the experimental station is clay-vertisol type. The available on season moisture was 234 and 280 mm for 2015–2016 and 2016–2017, respectively, during the growing season, whereas the average daily temperature was 14.1 °C for the first year and 13.5 °C for the second year. The heat-stress treatment followed the same agronomic practices, with the difference that at the time of booting (Z45) a 10 m^2 and 1.5 m high plastic tunnel was placed over each block (Figure 1) and left there until early dough stage (Z83). An electronic thermometer (temperature data logger) was placed in the middle of each block (normal and heat stressed) to reveal that the temperatures were up to 16° C higher inside the plastic tunnels, to reach a maximum of 49 °C (Figure 1). Marchouch is a drought prone site, and no rainfall occurred after Z45 in any of the two field seasons.

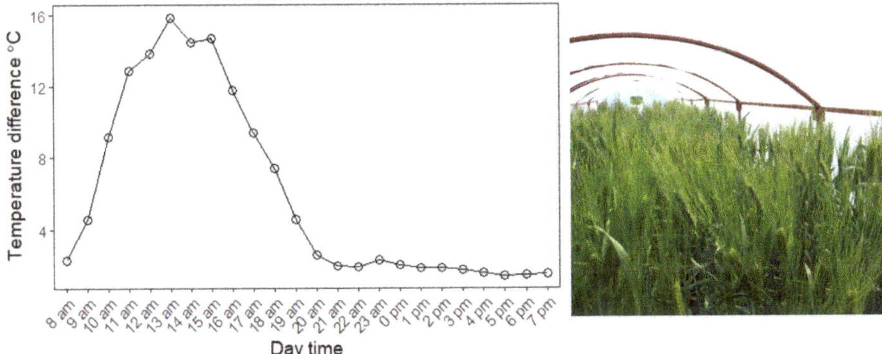

Figure 1. Mean temperature difference of 18 days over two seasons between the plastic tunnel-mediated heat stress and normal field conditions between 8 a.m. and 8 p.m., and a picture of the plastic tunnel at 9 a.m.

The following traits were recorded: days to heading (DTH) measured at the moment when the awns became visible, plant height (PH) measured from the ground to the top of the highest spike excluding the awns, and the number of fertile spikes per meter square (Spkm^2) was counted in a 0.25 m^2 area. The whole plot was harvested by hand and the dry biomass (Biom) was weighed before threshing. Grain yield (GY) was weighed for each plot and expressed as kg ha^{-1}. The weight of a thousand kernels (TKW) was expressed in grams. The harvest index (HI) was calculated as the ratio between GY and Biom. The grain number per spike (GNSpk) was derived from dividing grain number per meter square by Spkm^2 as follows:

$$\text{Grain number}/\text{m}^2 = \frac{\text{Grain weight of the plot}}{1.5\text{m}^2 \times \frac{\text{TKW}}{1000}} \tag{1}$$

$$\text{GNSpk} = \frac{\text{Grain number}/\text{m}^2}{\text{Spkm}^2} \tag{2}$$

The second and third sets were field tested in Kaedi, Mauritania (16°14″ N; 13°46″ W) during season 2014–2015 and 2015–2016 where the temperature reached a maximum of 41 °C and an average maximum daily temperature of 34 °C throughout the season. The trial was carried out under augmented design with a plot surface of 4.5 m^2. Standard agronomic management practices were adopted. Full details for this experiment are published elsewhere [35].

2.3. Data Analysis

A mixed linear model was run using the lme4 package [36] in R [37] to obtain best linear unbiased estimates (BLUEs) of the normally distributed traits. For count traits (DTH, Spkm^2, GNSpk), the generalized mixed linear model was used to get the BLUEs by Proc GLIMMIX in SAS. In both models, genotype, treatment, year, and replication were considered as fixed effects and block as random effect nested in treatment and year. Broad-sense heritability was calculated based on variance components from random model using the method suggested by DeLacy et al. [38]:

$$H^2 = \frac{\sigma^2 g}{\sigma^2 g + \frac{\sigma^2 GxT}{t} + \frac{\sigma^2 GxY}{y} + \frac{\sigma^2 GxTxY}{ty} + \frac{\sigma^2 e}{tyr}} \tag{3}$$

where: σ^2_{GxT} = genotype × treatment variance, σ^2_{GxY} = genotype × year variance, σ^2_{GxYxT} = genotype × treatment × year variance, σ^2_e = residual variance, r is the number of replications per treatment, t is the number of treatments, and y is the number of years.

Box-and-whisker plots where constructed by ggplot2 package [39] using the BLUEs combined over year per each treatment. The relationship between the target trait GY and yield components (GNSpk, TKW, Biom, HI) was studied using the Pearson correlation coefficient and the additive regression model. The critical value of the correlation significance was determined at 0.30 for $p < 0.05$ and 0.39 for $p < 0.01$ for 40 df using the corrplot package [40]. The additive model incorporates flexible forms (i.e., splines) of the functions to account for non-linear relationship contrary to linear regression model estimated via ordinary least squares [41]. For the additive model, the effective degree of freedom term determines the nature of the relationship between the predictor and the response variables where EDF = 1 indicates linearity and EDF > 1 the non-linearity. The additive regression analysis was performed using the mgcv package [42].

Two stress tolerance indices were calculated to identify the heat tolerant genotypes. The stress susceptibility index (SSI) [43,44] was calculated as follows:

$$SSI = \frac{[1 - (Ys)/(Yp)]}{[1 - (\tilde{Y}s)/(\tilde{Y}p)]} \tag{4}$$

where Ys and Yp are yield values of the genotypes evaluated under heat stress and normal conditions, respectively, and $\bar{Y}$s and $\bar{Y}$p are the mean yields of the lines evaluated under heat stress and normal conditions, respectively.

The stress tolerance (TOL) [45] was calculated as follows:

$$TOL = Yp - Ys. \tag{5}$$

The classInt package [46] was used to identify the possible number of class intervals of the indices for the frequency distribution of the subset.

The cut-off value for tolerant vs. susceptible genotypes for SSI was equal to 1, with lines having SSI < 1 being stress tolerant. Regarding the TOL index, the smaller TOL values indicate the genotypes with low yield depression and hence more tolerant. The experiment-wide TOL mean (1608 kg ha^{-1}) was identified as the cut-off value for tolerant vs. susceptible. The emmeans package [47] based on ANOVA model was used to discriminate among the grain yield means of haplotypes.

2.4. Genotyping and Marker-Trait Associations

Details of the genotyping step of the core set and panel have been previously discussed in Kabbaj et al. [34] and Sall et al. [35]. Briefly, 7652 high-fidelity polymorphic single nucleotide polymorphism (SNPs) were obtained, showing less than 1% missing data, minor allele frequency (MAF) higher than 5%, and heterozygosity less than 5%. The sequences of these markers were aligned with a cut-off of 98% identity to the durum wheat reference genome [48] (available at: http://www.interomics.eu/durum-wheat-genome), to reveal their physical position. The average length of the Axiom probe is of 75 bp, hence the 2% allowed miss-match was set to account for the existence of 1 SNP within each sequence. A sub-set of 500 highly polymorphic SNPs were selected on the basis of even spread along the genome, and used to identify the existence of population sub-structure, which revealed the existence of 10 main sub-groups [34]. To avoid bias, these 500 markers were then removed from all downstream association analysis. Linkage disequilibrium was calculated as squared allele frequency correlations (r^2) in TASSEL V 5.0 software [49], using the Mb position of the markers along the bread wheat reference genome. Linkage disequilibrium (LD) decay was estimated and plotted using the "Neanderthal" method [50]. The LD decay was measured at 51.3 Mb for $r^2 < 0.2$ as presented in Bassi et al. [51].

The genome wide association study (GWAS) was based on BLUEs of all the traits that displayed a significant treatment effect and the two stress tolerance indices. Two models were fitted and compared using two covariate parameters, Q (population structure) and K (Kinship). Q model was performed using a general linear model (GLM), and Q + K model using a mixed linear model (MLM). The best model for each trait was selected based on the quantile-quantile (Q-Q) plots [52]. Flowering time (DTH) was used as covariate in all analyses to remove the strong effects of flowering genes from the study. The value calculated for the LD decay of 51.3 Mb indicated that this association panel interrogated the 12,000 Mb of the durum wheat genome via 248 "loci hypothesis," and hence the Bonferroni correction for this panel was set to 3.1 LOD for $p < 0.05$ as suggested by Duggal et al. [53]. Local LD decay for $r^2 < 0.2$ was calculated for a 100 Mbp window around the marker with highest LOD for all marker-trait associations (MTAs) identified at a distance inferior to 104 Mbp (twice the LD decay). The MTAs that occurred at a distance inferior to twice the local LD were considered to belong to the same QTL. QTL associated to flowering time were removed from all downstream analyses (Table S2). A regression analysis was performed between the haplotype of the peak marker of each QTL to determine possible duplicate or homeolog loci. In addition, all the MTAs analyses were performed using Tassel 5 software [49].

2.5. Markers Conversion to KASP (Kompetitive Allele Specific PCR)

The array sequences of 20 markers associated to traits (MTA) were submitted to LGC Genomics for in-silico design of KASP primers using their proprietary software. Those that passed the in-silico criteria were purchased and used to genotype the independent validation set. For each marker that amplified and showed polymorphism, the regression cut-off between phenotype and haplotype was imposed at $r = 0.105$ following Pearson's critical value [54]. KASP markers AX-95260810, AX-94432276, and AX-95182463 were tested for association with grain yield, while AX-94408589 for association with biomass. In addition, the top 20 and worst 20 lines were considered as the true positive and true negative for heat tolerance. Hence, the accuracy was calculated as the ratio of the correct allelic call among all, sensitivity as the ratio of the correct positive allelic among the top 20 yielding lines, and specificity as the ratio of the correct negative (wt) allelic calls among the 20 worst yielding lines. The sequence of the validated KASP markers is provided in Table S3, or the primers can be ordered directly at LGC Genomics indicating the Axiom code used in this article.

3. Results

3.1. Agronomic Performance of the Genotypes and Sensitivity of Traits to Heat Stress

The combined analysis of variance across four environments (two different temperature treatments over two crop seasons) revealed significant genotypic differences for all traits measured (Table 1). The yield performance of the genotypes across environments averaged 2171 kg ha^{-1} and ranged from 352 kg ha^{-1} obtained under heat stress conditions for the lowest yielding line DWAyT-0215, to 4658 kg ha^{-1} under normal conditions for the highest yielding line DWAyT-0217. The top yielding line under heat-stress was the ICARDA/Moroccan cultivar 'Faraj' with an average yield of 2249 kg ha^{-1} over the two seasons.

Table 1. Descriptive statistics, component of trait variation, and heritability (h^2) among a set of 42 durum genotypes (G) tested under two treatments (T): normal and plastic tunnel-mediated heat stress during seasons 2015–2016 and 2016–2017.

Trait	Acronym	Mean	Min	Max	Genetic Variance (%)	Treatment Variance (%)	G × T (%)	h^2
Days to heading	DTH	92	71	109	34 **	1^{ns}	1^{ns}	0.78
Plant height (cm)	PH	81	71	92	60 **	1^{ns}	16^{ns}	0.76
Biomass (kg ha^{-1})	Biom	8407	4792	13,108	49 **	7 **	7 **	0.79
Spikes number per m^2	Spkm^2	524	370	640	14 **	1^{ns}	2 **	0.50
Grain yield (kg ha^{-1})	GY	2171	352	4658	30 **	44 **	12 *	0.63
Harvest index (%)	HI	26	1	50	15 **	34 **	13^{ns}	0.20
Thousand kernel weight (g)	TKW	36	27	45	48 **	1^{ns}	18 **	0.72
Grain number per spike	GNSpk	13	3	24	19 *	29 **	16 **	0.46

*, ** Significant at the 0.05 and 0.01 probability levels, respectively.

The treatment effect was significant only for Biom, GY, HI, and GNSpk, whereas DTH, PH, Spkm^2, and TKW were not significantly affected by treatments (Figure 2). The yield components were all significantly reduced under heat stress except TKW that showed a slight increase for the genotypes exposed to heat. The genotypes tested under plastic-tunnels had 61%, 54%, 42%, and 17% lower average GNSpk, GY, HI and Biom, respectively, compared to control. Relatively high heritability was observed for all the phenological and agronomical traits except for HI that had the lowest heritability ($h^2 = 0.20$).

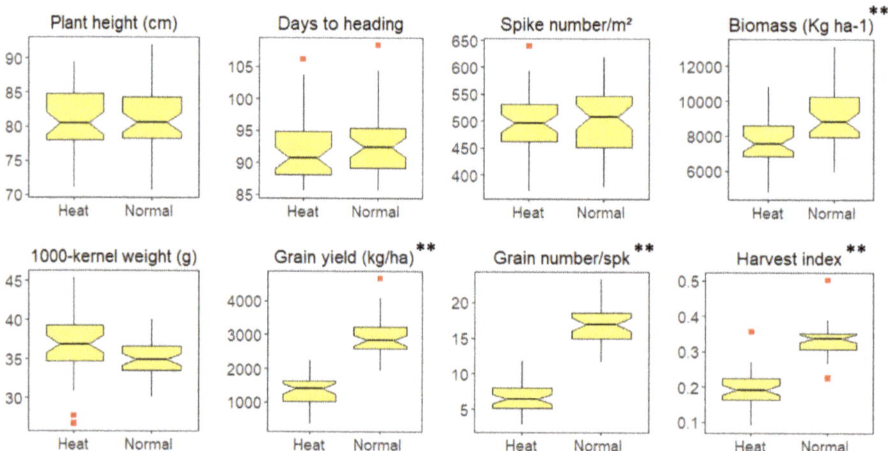

Figure 2. Boxplot of the best linear unbiased estimates (BLUEs) for various traits under two different environmental conditions (Heat: plastic tunnel-mediated heat stress and Normal) across two years. ** indicate significant difference between the means of control and heat-stressed plants at $p < 0.05$.

3.2. The Traits Interrelationship under Each Environmental Condition

Correlation analysis (Figure 3; Tables S4 and S5) was first conducted to investigate the interrelationship among all agronomic traits. Under both treatments, GNSpk had the highest association with GY (r = 0.81 under heat, r = 0.67 under normal), while Spkm^2 and TKW were the least correlated with GY. Biomass was also correlated with GY with r = 0.61 under heat and r = 0.67 under normal conditions. HI also showed a significant positive correlation with yield under both treatments, but its effect was stronger under heat stress (r = 0.72) than normal conditions (r = 0.54). DTH was not significantly correlated to any trait except HI (r = −0.44) under normal conditions.

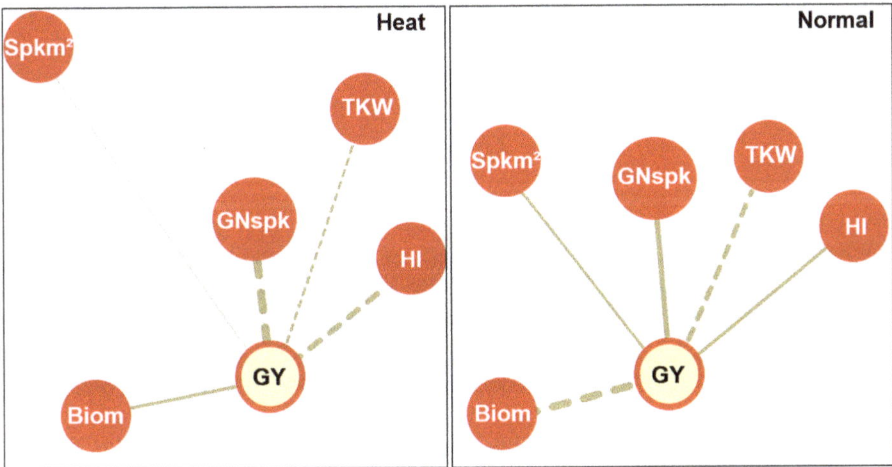

Figure 3. Relationships between grain yield (GY) and yield components (grain number per spike (GNspk), harvest index (HI), dry biomass (Biom), number of fertile spikes per meter square (Spkm^2), weight of a thousand kernels (TKW)) under plastic tunnel-mediated heat stress and normal conditions assessed by Pearson correlation and simple generalized additive model. The continuous grey line represents a linear relationship; the dashed grey line represents a non-linear relationship. The thickness of the line indicates the level of predictivity of the trait for GY. The length of the lines represents the correlation, the shorter the line the more the trait is correlated to GY.

Among yield components, the only significant and positive associations under the two environmental conditions were observed between Spkm^2, TKW, and Biom and between HI and GNSpk. Under heat conditions, a positive and significant correlation was noticed between GNSpk and Biom while under normal conditions HI was positively associated to TKW (Figure 3; Table S4).

The additive model was then used to further determine the nature of the relationship between GY and each predictor variable under normal and heat conditions (Figure 3; Table S5). The similarities observed between the two treatments in terms of the nature of relationship between GY and each of the predictors were the constantly linear and non-linear relationship between Spkm^2, TKW and the response variable GY, respectively.

GNSpk was considered the best predictor (deviance = 0.73%) with a complex relationship (EDF = 2.64) with GY under heat stress, whereas under normal conditions this trait was the second best predictor (deviance = 0.44%) with a linear relationship (EDF = 1). A similar trend was observed for HI in both treatments. Biom was found to be the best predictor (deviance = 0.52%) for GY with a non-linear relationship (EDF = 2.52) under normal conditions (Table S2; Figure S1).

3.3. Stress Tolerance Indices

Two different stress tolerance indices were calculated for GY: SSI and TOL (Figure 4). The genotypes showed wide variation for these indices. Seven SSI groups were identified with four having an SSI lower than 1 and the three remaining groups of genotypes having SSI > 1. The frequency distribution of the panel showed a wide variation and indicated the presence of susceptibility, with 45% of the genotypes falling in the very heat-susceptible class of SSI higher than 1, and only 7% of the lines showing high tolerance at SSI < 1. For TOL index, seven groups were also identified with 48% of the lines showing high yield depression and 5% of the genotypes presenting high stability. The smaller TOL values indicate the genotypes with low yield depression and hence more tolerant. However, good heat tolerance can also be reached by low yielding lines, but their value for breeding would be questionable. Hence, a scatterplot was devised to compare the GY under normal conditions and each of the heat indices (SSI and TOL). Five genotypes (four ICARDA lines, one Moroccan cultivar): Kunmiki, Berghouata1, Margherita2, IDON37-141, and Ourgh were found to have above average yield, low yield depression (low TOL values) and good heat tolerance (SSI < 1).

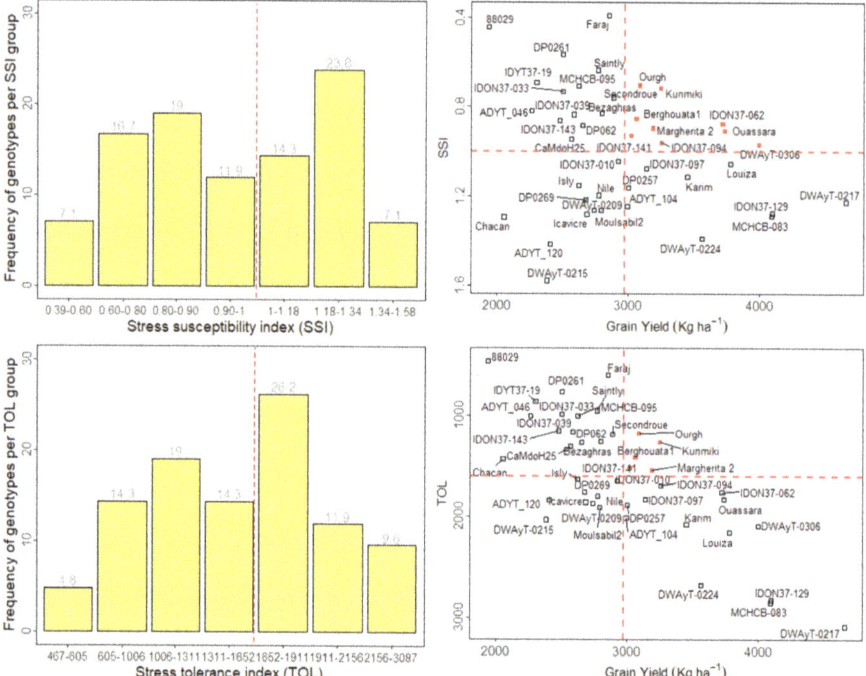

Figure 4. Two different stress tolerance indices SSI (stress susceptibility index) and TOL (tolerance index) of grain yield, comparing plastic tunnel-mediated heat stress with normal conditions for the 42 durum wheat genotypes. The bars plot shows the frequency distribution of SSI and TOL for the genotypes tested. The dashed red lines mark the separation between tolerant (left) and susceptible (right) genotypes. The scatter plot shows the yield performance of genotypes tested under normal conditions against each of SSI and TOL. The vertical dashed red lines indicate the average GY. The horizontal dashed red lines indicate the cut-off value for tolerant vs. susceptible genotypes for each index. Red dots indicate genotypes that were identified as superior by both bi-plots.

3.4. Markers Associated to Heat Stress Tolerance

A total of 204 MTAs were identified for four traits (GY, GNSpk, HI and Biom) under both stress and normal conditions and 49 MTAs were recorded for the two GY stress tolerance indices. Regression analysis and clustering based on local LD decay confirmed that these associations were distributed over 12 loci (Table 2 and Table S6). Chromosome 1A had the highest number of MTAs (27) while chromosome 4A had the lowest (6).

Under normal conditions, 56 MTAs were detected for three traits GY, GNSpk, and HI, with the third trait having the highest number of MTAs (48). No common region for these traits was identified under the non-stress environment. Under heat stress, a higher number of associations (148) were identified with trait variation (r^2) ranging from 0.25 to 0.36. The highest number of MTAs were detected for GNSpk distributed over 10 different loci, followed by HI on six loci. A common region for GY, GNSpk, HI, and Biom was identified under the heat condition on chromosome 6BS. Loci associated with both GNSpk and HI were detected on 1AL, 1BL, 2AL, 3AL, and 3BL. For heat tolerance indices (SSI-GY and TOL-GY), 49 MTAs were identified. The common loci associated with the two indices were on chromosomes 2AL, 5AL, and 5BL, while the loci on chromosomes 1AL and 6BS were identified only for TOL-GY and SSI-GY, respectively.

A comparison of the significant loci under each treatment and including the heat tolerance indices indicated a locus on chromosome 2AL, which was consistently identified for the indices, and both treatments for GNSpk and HI. Two loci on chromosomes 3AL and 3BL were associated with GNSpk and HI under both control and stress conditions, but were not associated with any of the indices. Three significant loci on chromosomes 1AL, 5BL, and 6BS were shared among heat stress treatment and stress tolerance indices, but not under normal conditions, making of these the most interesting genomic regions that specifically respond to heat stress. Overall, a total of 12 unique significant loci were identified (numbered QTL.ICD.Heat.01–QTL.ICD.Heat.12) and can be consulted in Table 2. Local LD decay was estimated for the 100 Mbp genomic region surrounding the peak marker. It varied between 31.7 and 108.7 Mbp, or a −38% to 112% variation compared to the average LD decay calculated for the whole panel (51.3 Mbp). This variation was accounted for to determine the correct physical size in each genomic region to assign multiple MTAs to the same QTL.

Table 2. Quantitative trait loci (QTLs) associated with multiple traits under plastic tunnel-mediated heat stress, normal conditions, and based on stress indices.

Locus	Trait	Chr. [†]	Main Marker	Position [‡] (bp)	Local LD (Mbp)	Max LOD	Max r^2	Heat Stress	Normal	Indices
QTL.ICD.Heat.01	GNspk, HI, TOL-GY	1AL	AX-94863732	570,040,339	31.7	3.38	0.27	*		*
QTL.ICD.Heat.02	GNspk, HI	1BL	AX-94447402	632,403,981	43.1	3.38	0.27	*		
QTL.ICD.Heat.03	GNspk, HI, SSI-GY, TOL-GY	2AL	AX-94538070	748,624,588	36.3	3.06	0.25	*	*	*
QTL.ICD.Heat.04	GY, HI	2BS	AX-95193898	6,012,904	36.0	3.67	0.36		*	
QTL.ICD.Heat.05	GNspk, HI	3AL	AX-95632723	562,421,267	75.4	3.39	0.27	*	*	
QTL.ICD.Heat.06	GNspk, HI	3BL	AX-95174625	788,551,042	85.4	3.38	0.27	*	*	
QTL.ICD.Heat.07	GNspk	5AS	AX-95247611	27,923,949	108.7	3.38	0.27	*		
QTL.ICD.Heat.08 [§]	SSI-GY, TOL-GY	5AS	AX-94631521	421,078,546	41.3	4.93	0.45			*
QTL.ICD.Heat.09 [§]	GNspk, SSI-GY, TOL-GY	5BS	AX-95182463	427,098,066	50.3	4.17	0.37	*		*
QTL.ICD.Heat.10 [§]	GNspk, HI, Biom, SSI-GY	6BS	AX-94408589	157,777,006	56.0	3.20	0.36	*		*
QTL.ICD.Heat.11	GNspk	7AL	AX-95074729	660,833,752	153.6	3.60	0.29	*		
QTL.ICD.Heat.12	GNspk, HI	7AS	AX-94381852	16,943,364	44.8	3.42	0.37		*	

[†] Chr.—Chromosome, based on alignment to durum wheat genome assembly [48].*—Significant QTL; [‡]—Based on alignment to durum wheat genome assembly [48]; [§]—These QTLs have been converted into KASP markers and validated; GNspk—Grain number per spike; HI—Harvest index; TOL-GY—Tolerance index for grain yield; SSI-GY—Stress susceptibility index for grain yield; GY—Grain yield; Biom—Biomass.

3.5. Effect of Different Allele Combination on Yield Performance

The loci identified on chromosomes 1AL, 5BL, and 6BS appeared as the most critical for heat tolerance and were then tested further. These regions were associated with the control of multiple traits under heat stress: GY, GNspk, HI, Biom and the two indices SSI-GY and TOL-GY. A set of 208 modern lines were investigated for haplotype diversity at these three loci. Five groups with different allelic combinations were identified (Figure 5). Their allelic effect on GY was then assessed when field tested under high temperatures along the Senegal River [35]. The haplotype class with positive alleles at all three loci had the highest GY average reaching 2381 kg ha^{-1} with a maximum value of 3856 kg ha^{-1}. Genotypes of the haplotype classes with only two favorable alleles reached GY of 2199 and 2103 kg ha^{-1}, while lines that only carried one positive allele 2103 and 2023 kg ha^{-1} (Figure 5). ANOVA confirmed that the haplotype group with all three positive alleles was significantly superior to the others.

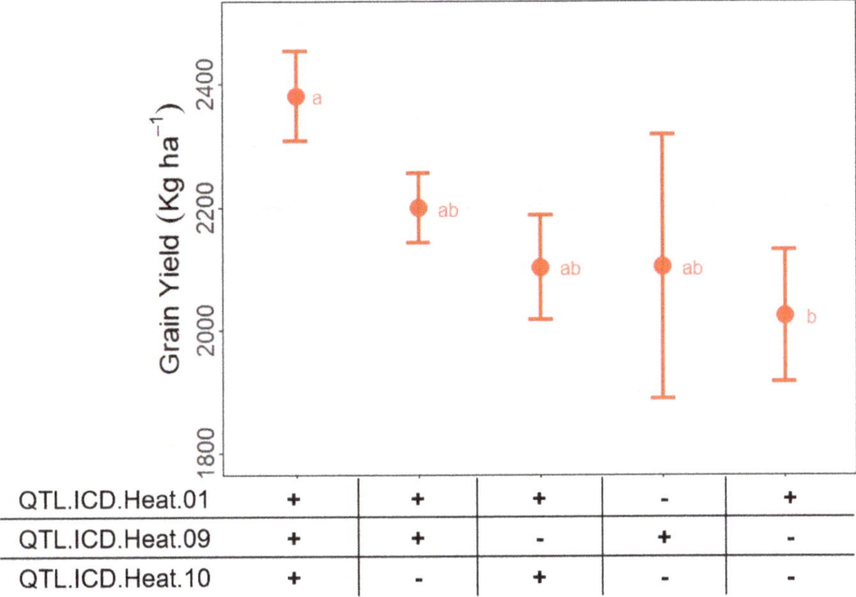

Figure 5. Effect of different allele combinations of the significant loci on yield performance of 208 accessions tested under heat stressed conditions along the Senegal River. The circle indicates the average of each class over 2 years, and the whiskers show the standard error of the mean. The accessions were divided into five clusters based on their haplotype for three major QTLs: "+" mark the positive and "-" the wild-type alleles. Letters (a, b, ab) indicate significant differences between the clusters.

3.6. Validation of Markers for Marker Assisted Selection

To effectively deploy in breeding the most interesting QTLs via MAS, it is first required a step of validation using more affordable marker methodologies and in different genetic backgrounds and environments. A total of 20 MTA sequences linked to important agronomical and spike fertility traits were submitted for KASP primers design. Among these, only 14 could be successfully designed, and 11 identified a polymorphism within the validation set. Four showed significant ($p < 0.05$) correlation to the test phenotype (Figure 6). Three QTLs were represented by these four markers, AX-95260810 and AX-94432276 tagged QTL.ICD.Heat.08 on chromosome 5AL, AX-95182463 underlines QTL.ICD.Heat.09 on chromosome 5BL, and AX-94408589 tags QTL.ICD.Heat.10 on chromosome 6BS. The latter two QTLs are among the three main effect regions identified in this study (Figure 5).

AX-95260810 reached 15% correlation to grain yield under heat, 74% accuracy, 43% sensitivity, and 100% specificity. Especially, its ability to identify 100% of non-heat tolerant entries is particularly remarkable. AX-95182463 and AX-94408589 also reached significant correlations of 14% and 32% for grain yield and biomass under severe heat, respectively, with sensitivities of 62% and 40%, accuracies of 30% and 65%, and specificities of 4% and 90%. Overall, AX-9526081 and AX-94408589 appeared as the most suitable for MAS application.

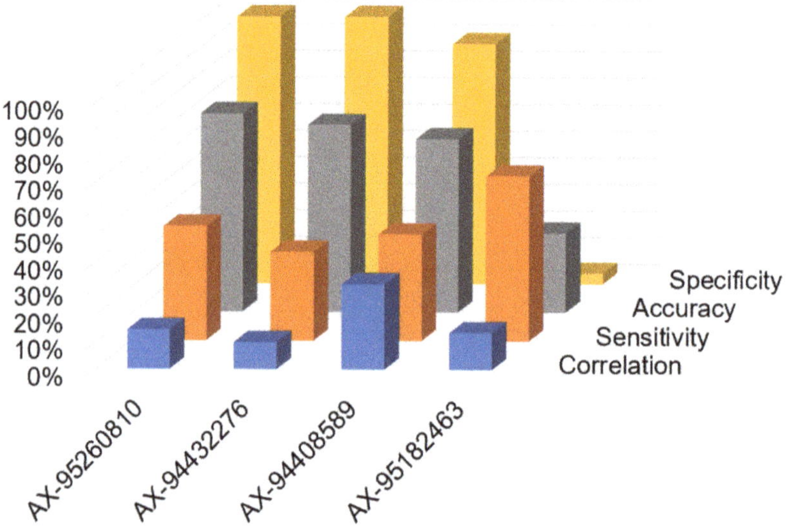

Figure 6. Kompetitive Allele Specific PCR (KASP) markers validation on an independent set of 94 elite lines of ICARDA tested under severe heat for grain yield and biomass. Correlation was measured between the BLUE for grain yield recorded along the Senegal River and the haplotype score. Accuracy, sensitivity, and specificity where determined using only the top 20 and worst 20 lines. AX-95260810 and AX-94432276 tag QTL.ICD.Heat.08, AX-95182463 tags QTL.ICD.Heat.09, AX-94408589 tags QTL.ICD.Heat.10.

4. Discussion

4.1. Evaluation of the Phenotypic Performance of Yield and Yield Components under Normal and Heat Stress Conditions

Several studies reported that wheat plants are very sensitive to elevated temperatures during flowering and grain filling phases [9,55,56], due to a reduction in seed development and fertility [56–58]. This study evaluated a set of durum wheat genotypes derived from a global collection for GY and yield components under heat and normal conditions. The genetic and phenotypic diversity shown by this set together with its relatively similar flowering time, promote it as an ideal panel to test heat tolerance. Further, the plastic tunnel method deployed here allowed to increase the temperatures well above 21 °C, the value that defines the absence of the stress [9]. A similar methodology was also successfully deployed by Corbellini et al. [54] to study the effect of heat shock proteins on technological quality characteristics. Compared to timely vs. delayed sowing experiments to simulate heat stress, the use of the plastic tunnel method avoids incurring false discovery due to changes in the phenological behavior of plants.

In the present study, a short and severe episode of heat stress was applied from the beginning of heading to the early dough stage, and resulted in 54% reduction in grain yield. This was in agreement with the study conducted by Ugarte et al. [59] that found a reduction of up to 52% when thermal treatment was applied via transparent chambers. Interestingly, our stress treatment caused an average temperature increase of 10 °C, which caused an average GY reduction of 5.4% for each 1 °C raise. This value is well within the 4.1% to 6.4% interval suggested by Liu et al. [10] for 1 °C raise in temperatures. GNSpk was the most affected trait (−61%) with the highest positive correlation to GY. This is in good agreement with previous studies that have shown that seed setting is the most sensitive parameter to heat stress, with a noticeable influence on yield [28,60–62]. Still, its non-linear relationship to yield confirms the complexity of the trait. Biom and HI were also found to have an influence on yield [63,64] with different relationships based on the occurrence of the stress. The presence of dissimilarities of the associations between the two treatments indicates clearly that there is a trade-off among the yield components as previously reported by Sukumaran et al. [65] for grain weight and grain number. Variation of one of the yield components affect the others positively or negatively. Compared to the simple regression, the additive model allowed to reveal the complexity of the relationship between GY and yield related traits.

The stress index SSI was developed by Fisher and Maurer [43] and modified by Nachit and Ouassou [44] as a useful indicator and a good parameter for selection. It measures the severity of the heat stress [66,67] and was also used in earlier studies in wheat to seek heat tolerant genotypes [23,68,69]. The TOL index is instead useful for selecting against yield depression, and it was used in several studies for heat or drought tolerance in wheat [27,44,67,70]. Improving heat tolerance should not be based on the use of these criterions alone as was suggested by Clarke et al. [71]. It is important to select simultaneously for good yield performance coupled with good adaptability (SSI < 1) and stability (low TOL) [44]. In that sense, the accessions Kunmiki, Berghouata1, Margherita2, and IDON37-141 originated from ICARDA durum wheat program, and Ourgh, a Moroccan cultivar, have been identified as high yielding genotypes that also show good heat stress tolerance based on the two indices.

4.2. Dissection of Heat-Specific QTLs Associated with Yield-Related Traits and Stress Tolerance Indices

The significant correlation identified between yield and its components were not linear in nature, and tend to change their mode of action based on the occurrence of the stress. Therefore, several physiological processes are simultaneously involved in protecting the wheat plant from the heat stress [72], and there is value in dissecting it into its genetic components. In this study GWAS was used to identify the genetic regions controlling the response of the various traits. To prevent the confounding effect that phenology-related loci might have [73], MTAs were identified for DTH and removed from downstream analysis. Additionally, flowering time was used as covariate in all analyses for the other traits. Very few MTAs for DTH were observed either in normal or stressed conditions due to the synchronized flowering of the entries used in this study. This indicated the absence of confounding effects between the two trials. i.e., almost all the accessions were exposed to the same conditions in each developmental phase [74] before imposing the stress.

Out of 12 QTLs identified, three occurred only when the heat stress was imposed, including indices. These three main genomic regions occurred on chromosomes 1AL, 5BL, and 6BS, and were considered as QTLs controlling heat tolerance. These three loci were confirmed by mean of haplotype analysis on a larger panel of modern lines (208 entries) field tested under severe heat along the Senegal River valley [35], to confirm that the presence of the positive alleles at all three loci provided a significant GY advantage of +182 kg ha^{-1} (+8%). The QTL on the long arm of chromosome 1A controlled GNSpk, HI, and TOL-GY, and it explained up to 27% of the phenotypic variation. In a study with double haploid population of bread wheat, Heidari et al. [75] identified a major QTL on the same chromosome (1A), influencing grain number per spike, grain weight per spike, and spikes/m^2. However, their phenotypic assessment was not performed under heat stress, the marker systems used was different compared to our study and the locus was identified in the short arm of chromosome 1A. Therefore, it is quite difficult to align the results from that study to the current one. Another study had previously reported many MTAs on chromosome 1A detected for yield components under heat stress, but all were found to have a pleiotropic relationship with days to heading and were also located on the short arm of 1A [26], instead of 1AL found here. A heat-specific QTL was also detected on the same chromosome in the short arm for spikelet compactness and leaf rolling in bread wheat [76]. An earlier study identified a QTL on 1AS for yield but associated with different stress conditions [77]. To the best of our knowledge, this is the first time that this region on 1AL is presented as associated to GNSpk, HI, and TOL-GY in durum wheat under heat stress conditions. The second major QTL region was detected on the long arm of chromosome 5B and found to be associated with GNSpk and the two indices SSI-GY and TOL-GY, contributing to 37% of the phenotypic variation. A region in the short arm of the same chromosome has been previously reported to be associated with grain number per square meter in bread wheat [76], and controlling thousand grain weight in durum wheat [27] under combined drought and heat stress. Shirdelmoghanloo et al. [25] and Acuna-Galindo et al. [78] reported loci for grain weight and other important traits on chromosome 5B under heat and non-heat conditions in hexaploid wheat. On the other hand, the same chromosome has been previously suggested to carry heat-specific QTLs for yield per se in bread wheat [26]. Sukumaran et al. [27] identified markers for heat susceptibility (HSI or SSI) and tolerance (TOL) indices for yield and grain number per square meter on the short arm of the chromosome 5B. Mason et al. [64] also detected QTL for HSI for kernel number on 5BL in bread wheat. The genomic region identified in this study on 5BL is likely to be a new QTL since no information has been reported earlier for this locus associated to GNSpk, SSI-GY, and TOL-GY in durum wheat and specific to heat stress, but we cannot exclude that it overlaps with previously reported QTLs. A third heat-responsive locus was identified on the short arm of chromosome 6B related to GY, SSI-GY, GNspk, HI, and Biom accounting for 36% of the phenotypic variance. An earlier study on bread wheat identified a locus on chromosome 6BS underpinning chlorophyll loss rates and heat susceptibility index for grain weight and chlorophyll loss rates under heat-stress conditions [25]. Under post-anthesis high temperatures stress, Vijayalakshmi et al. [20] reported a QTL on the short arm of chromosome 6B for senescence related traits in hexaploid wheat. McIntyre et al. [79] and Pinto et al. [21] reported QTLs on chromosome 6BL that were associated with many important traits (grain number per square meter and grain yield and water-soluble carbohydrate content) related to drought and heat tolerance. Ogbonnaya et al. [26] found a locus on the short arm of chromosome 6B for grain yield under heat stress in bread wheat. These previously reported QTLs in 6B could overlap with the one identified in this study, but they were either identified not in association with heat tolerance or detected in hexaploidy wheat. Therefore, this region is also assumed to have been reported for the first time here in relationship to heat tolerance for durum wheat. This locus affects multiple traits (GY, GNspk, HI, Biom, and two heat susceptibility indexes) and hence it is of good importance for deployment in breeding. The principal breeding objective is to develop varieties with high grain yield and stability when exposed to different stresses. However, grain yield is a complex trait controlled by many genes and strongly influenced by the environment [80–86]. Therefore, a good understanding of traits and underlying loci associated with tolerance to elevated temperatures is of a great importance for breeding new heat tolerant cultivars [87].

4.3. Pyramiding Heat-Tolerant QTLS via MAS

Three loci on chromosomes 1AL, 5BS, and 6BS showed an additive nature by means of haplotype analysis (Figure 5), revealing that only the combination of all three positive alleles generated a true yield advantage. Among the most heat tolerant elite lines identified here 'Kunmiki', 'Berghouata1', and 'Ourgh' confirmed to harbor the positive alleles for all three loci. This prompts their use in crossing schemes to pyramid the positive alleles, as well as the deployment of simple marker system to conduct MAS.

Axiom to KASP marker conversion and validation was attempted for 20 MTAs. Eleven KASP markers generated polymorphic haplotypes in an independent set of ICARDA elite lines. Four revealed a significant ($p < 0.05$) correlation to GY and biomass assessed under severe heat along the Senegal River Valley (Figure 6). In particular, AX-95182463 tags QTL.ICD.Heat.09 located on chromosome 5B and it revealed good correlation and sensitivity, but lacks in accuracy and specificity, and it is hence protected from Type II errors, but prone to Type I, with several elite lines wrongly identified as carrying the positive alleles. AX-95260810 tags QTL.ICD.Heat.08, linked to the two stress tolerance indices for GY (SSI-GY and TOL-GY) located on chromosome 5A. AX-94408589 tags QTL.ICD.Heat.10 located on chromosome 6B, and associated to several traits GNspk, HI, Biom, SSI-GY. In these two cases, the KASP markers explained 15% and 33% of the phenotypic variation of an independent validation set, with 100% and 90% specificity, and 74% and 65% accuracy, but medium sensitivity (43% and 40%). As such, these markers are protected against Type I errors (no false positive), but prone to Type II errors, with several elite lines identified as not carrying the positive allele while instead being tolerant to heat. Hence, while all converted KASP markers are prone to different types of errors, these three markers can be considered as validated and ready to be deployed in breeding. The combination of the three might represent a more stringent approach to protect against both types of errors. An additional nine QTLs were identified in this study, and their KASP conversion and validation are still ongoing and will require better targeted efforts to be achieved.

5. Conclusions

Heat stress causes a complex cascade of negative effects on the wheat plant, resulting in drastic reductions in grain yield. The deployment of heat tolerant varieties that will benefit greatly farmers requires first to enhance our understanding of this mechanism and loci governing it. Our study combined a discovery phase with a core set tested over two field seasons in Morocco under artificial heat-treatment with plastic tunnels, followed by a different confirmation set of germplasm grown for two seasons in Kaedi, Mauritania under severe natural heat, and completed with one final validation set tested one season in Kaedi. Our results confirmed that spike fertility (GNSpk) and maintenance of green leaves (Biom) are the most critical traits to drive tolerance to this stress, and hence should be the primary targets of durum wheat breeders. Further, the deployment of plastic tunnels proved to be a strategic methodology to study this stress and reveal its mechanisms without affecting the phenology of the plant. In addition, 12 loci were identified as responsible for controlling the main heat tolerance traits. Among these, three were activated only when the stress occurred and hence represent ideal targets for breeding. Two of these were validated into a KASP marker and are now ready for deployment via MAS, especially if associated with a third, also validated, KASP. Finally, three ICARDA elite lines and one Moroccan cultivar were confirmed as tolerant to heat, with high grain yield, and carrying positive alleles for three main QTLs. These are freely available and should be incorporated as crossing parents by other breeding programs. Altogether, this study has confirmed the key traits for heat tolerance as well as a new methodology to study it in durum wheat, it has revealed the main loci controlling these traits and proceeded to validate three of them for MAS, and it has also provided freely available elite lines to breed new cultivars better adapted to the stress.

Supplementary Materials: Table S1: List of durum wheat genotypes evaluated under plastic tunnel-mediated heat stress in the present study, Table S2: Markers associated with days to heading (DTH) under heat stress and normal conditions, Table S3: Sequence information of the KASP markers, Table S4: Pearson correlation matrix between all the measured traits under heat conditions (upper part) and normal (lower part) conditions. GY—Grain yield; Biom—Biomass; HI—Harvest index; Spkm^2—Spikes per square meter: GNspk—Grain number per spike; TKW—Thousand kernel weight; DTH—Days to heading. *, ** Significant at the 0.05 and 0.01 probability levels, respectively, Table S5: Correlation (r), linear regression estimated via ordinary least squares (OLS) and flexible regression estimated via regression additive model. (**a**) Under heat stress. (**b**) Under normal conditions, Table S6: Regression matrix between the haplotype of the peak markers for the 13 identified QTLs. *, significant loci similarity at $p < 0.05$ consistent with homeologous relationship; **, significant loci identity ($p < 0.01$) consistent with wrongly assigned genomic position, Figure S1: Plots of the additive regression model showing GNspk, biom, TKW, spkm^2 and HI as the spline function of the target trait grain yield (GY). (**a**) Under heat stress. (**b**) Under normal conditions

Author Contributions: F.M.B., M.N. and K.E.H. conceived and designed the study. K.E.H. and F.M.B. performed the field experiment. A.T.S. performed the field experiment in the Senegal river. A.A. contributed in the genotyping. K.E.H. and F.M.B. analyzed the data. K.E.H. Wrote the original draft. K.E.H., B.B., A.F.M., A.A., M.N., and F.M.B. wrote or reviewed the manuscript. All authors read and approved the final manuscript.

Funding: This study was funded by the Australian Grains Research and Development Corporation (GRDC) project ICA00012: Focused improvement of ICARDA/Australia durum germplasm for abiotic tolerance, while the field work along the Senegal River was funded by the Swedish Research Council (Vetenskapsradet) U-Forsk2013 project 2013-6500, "Deployment of molecular durum breeding to the Senegal Basin: capacity building to face global warming" and U-Forsk2018 project 2017-05522, "Genomic prediction to deliver heat tolerant wheat to the Senegal River basin: phase II." The marker conversion work was covered by the International Treaty on Plant Genetic Resources for Food and Agriculture 2014-2015-2B-PR-02-Jordan: "An Integrated Approach to Identify and Characterize Climate Resilient Wheat for the West Asia and North Africa."

Acknowledgments: The authors wish to acknowledge the technical assistance provided by A. Rached and all ICARDA durum wheat program staff in handling field activities.

Conflicts of Interest: The authors declare no conflict of interest.

References

1. Hays, D.; Mason, E.; Hwa, D.J.; Menz, M.; Reynolds, M. Expression quantitative trait loci mapping heat tolerance during reproductive development in wheat (*T. aestivum* L.). In *Wheat Production in Stressed Environment*; Buck, H.T., Nisi, J.E., Salomon, N., Eds.; Springer: Amsterdam, The Netherlands, 2007; pp. 373–382.

2. Saini, H.; Aspinall, D. Abnormal sporogenesis in wheat (*Triticum aestivum* L.) induced by short periods of high temperature. *Ann. Bot.* **1982**, *49*, 835–846. [CrossRef]

3. Plaut, Z.; Butow, B.J.; Blumenthal, C.S.; Wrigley, C.W. Transport of dry matter into developing wheat kernels and its contribution to grain yield under post-anthesis water deficit and elevated temperature. *Field Crops Res.* **2004**, *86*, 185–198. [CrossRef]

4. Hunt, L.A.; van der Poorten, G.; Pararajasingham, S. Post-anthesis temperature effects on duration and rate of grain filling in some winter and spring wheats. *Can. J. Plant Sci.* **1991**, *71*, 609–617. [CrossRef]

5. Wollenweber, B.; Porter, J.R.; Schellberg, J. Lack of interaction between extreme high-temperature events at vegetative and reproductive growth stages in wheat. *J. Agron. Crop Sci.* **2003**, *189*, 142–150. [CrossRef]

6. Dias, A.S.; Lidon, F.C. Bread and durum wheat tolerance under heat stress: A synoptical overview. *J. Food Agric.* **2010**, *22*, 412–436. [CrossRef]

7. Dolferus, R.; Ji, X.; Richards, R.A. Abiotic stress and control of grain number in cereals. *Plant Sci.* **2011**, *181*, 331–341. [CrossRef] [PubMed]

8. Porter, J.R.; Gawith, M. Temperatures and the growth and development of wheat: A review. *Eur. J. Agron.* **1999**, *10*, 23–36. [CrossRef]

9. Farooq, M.; Bramley, H.; Palta, J.A.; Siddique, K.H.M. Heat stress in wheat during reproductive and grain-filling phases. *Crit. Rev. Plant Sci.* **2011**, *30*, 1–17. [CrossRef]

10. Liu, B.; Asseng, S.; Muller, C.; Ewert, F.; Elliott, J.; Lobell, D.B.; Rosenzweig, C. Similar estimates of temperature impacts on global wheat yield by three independent methods. *Nat. Clim. Chang.* **2016**, *6*, 1130–1136. [CrossRef]

11. Lillemo, M.; van Ginkel, M.; Trethowan, R.M.; Hernandez, E.; Crossa, J. Differential adaptation of CIMMYT bread wheat to global high temperature environments. *Crop Sci.* **2005**, *45*, 2443–2453. [CrossRef]

12. Hansen, J.; Sato, M.; Ruedy, R. Perception of climate change. *Proc. Natl. Acad. Sci. USA* **2012**, *109*, 2415–2423. [CrossRef] [PubMed]

13. United Nations Economic and Social Commission for Western Asia (ESCWA). *Arab Climate Change Assessment Report—Main Report*; E/ESCWA/SDPD/2017/RICCAR/Report; United Nations Publication: Beirut, Lebanon, 2017.

14. Ayeneh, A.; van Ginkel, M.; Reynolds, M.P.; Ammar, K. Comparison of leaf, spike, peduncle, and canopy temperature depression in wheat under heat stress. *Field Crops Res.* **2002**, *79*, 173–184. [CrossRef]

15. Yang, J.; Sears, R.G.; Gill, B.S.; Paulsen, G.M. Quantitative and molecular characterization of heat tolerance in hexaploid wheat. *Euphytica* **2002**, *126*, 275–282. [CrossRef]

16. Nachit, M.M.; Elouafi, I. Durum adaptation in the Mediterranean dryland breeding: Breeding, stress physiology, and molecular markers. *Crop Sci. Soc. Am. Am. Soc. Agron. CSSA* **2004**, *32*, 203–218.

17. Reynolds, M.P.; Hobbs, P.R.; Braun, H.J. Challenges to international wheat improvement. *J. Agric. Sci.* **2007**, *145*, 223–227. [CrossRef]

18. Tuberosa, R. Phenotyping for drought tolerance of crops in the genomics era. *Front. Physiol.* **2012**, *3*, 347. [CrossRef] [PubMed]

19. Reynolds, M.; Langridge, P. Physiological breeding. *Curr. Opin. Plant Biol.* **2016**, *31*, 162–171. [CrossRef] [PubMed]

20. Vijayalakshmi, K.; Fritz, A.K.; Paulsen, G.M.; Bai, G.; Pandravada, S.; Gill, B.S. Modeling and mapping QTL for senescence-related traits in winter wheat under high temperature. *Mol. Breed.* **2010**, *26*, 163–175. [CrossRef]

21. Pinto, R.S.; Reynolds, M.P.; Mathews, K.L.; McIntyre, C.L.; Olivares-Villegas, J.J.; Chapman, S.C. Heat and drought adaptive QTL in a wheat population designed to minimize confounding agronomic effects. *Theor. Appl. Genet.* **2010**, *121*, 1001–1021. [CrossRef]

22. Bennett, D.; Reynolds, M.; Mullan, D.; Izanloo, A.; Kuchel, H.; Langridge, P.; Schnurbusch, T. Detection of two major grain yield QTL in bread wheat (*Triticum aestivum* L.) under heat, drought and high yield potential environments. *Theor. Appl. Genet.* **2012**, *125*, 1473–1485. [CrossRef]

23. Paliwal, R.; Röder, M.S.; Kumar, U.; Srivastava, J.P.; Joshi, A.K. QTL mapping of terminal heat tolerance in hexaploid wheat (*T. aestivum* L.). *Theor. Appl. Genet.* **2012**, *125*, 561–575. [CrossRef] [PubMed]

24. Talukder, S.K.; Babar, M.A.; Vijayalakshmi, K.; Poland, J.; Prasad, P.V.V.; Bowden, R.; Fritz, A. Mapping QTL for the traits associated with heat tolerance in wheat (*Triticum aestivum* L.). *BMC Genet.* **2014**, *15*, 97. [CrossRef] [PubMed]

25. Shirdelmoghanloo, H.; Taylor, J.D.; Lohraseb, I.; Rabie, H.; Brien, C.; Timmins, A.; Martin, P.; Mather, D.E.; Emebiri, L.; Collins, N.C. A QTL on the short arm of wheat (*Triticum aestivum* L.) chromosome 3B affects the stability of grain weight in plants exposed to a brief heat shock early in grain filling. *BMC Plant Biol.* **2016**, *16*, 100. [CrossRef] [PubMed]

26. Ogbonnaya, F.C.; Rasheed, A.; Okechukwu, E.C.; Jighly, A.; Makdis, F.; Wuletaw, T.; Hagras, A.; Uguru, M.I.; Agbo, C.U. Genome-wide association study for agronomic and physiological traits in spring wheat evaluated in a range of heat prone environments. *Theor. Appl. Genet.* **2017**, *130*, 1819–1835. [CrossRef] [PubMed]

27. Sukumaran, S.; Reynolds, M.P.; Sansaloni, C. Genome-Wide Association Analyses Identify QTL Hotspots for Yield and Component Traits in Durum Wheat Grown under Yield Potential, Drought, and Heat Stress Environments. *Front. Plant Sci.* **2018**, *9*, 81. [CrossRef] [PubMed]

28. Sall, A.T.; Cisse, M.; Gueye, H.; Kabbaj, H.; Ndoye, I.; Maltouf, A.F.; El-Mourid, M.; Ortiz, R.; Bassi, F.M. Heat tolerance of durum wheat (*triticum durum* desf.) elite germplasm tested along the Senegal river. *J. Agric. Sci.* **2018**, *10*, 217–233. [CrossRef]

29. Pradhan, G.P.; Prasad, P.V.V.; Fritz, A.; Kirkham, M.B.; Gill, B.S. Effects of drought and high temperature stress on synthetic hexaploid wheat. *Funct. Plant Biol.* **2012**, *39*, 190–198. [CrossRef]

30. Weldearegay, D.F.; Yan, F.; Jiang, D.; Liu, F. Independent and combined effects of soil warming and drought stress during anthesis on seed set and grain yield in two spring wheat varieties. *J. Agron. Crop Sci.* **2012**, *198*, 245–253. [CrossRef]

31. Jha, U.C.; Bohra, A.; Singh, N.P. Heat stress in crop plants: Its nature, impacts and integrated breeding strategies to improve heat tolerance. *Plant Breed.* **2014**, *133*, 679–701. [CrossRef]

32. Mishra, S.C.; Singh, S.K.; Patil, R.; Bhusal, N.; Malik, A.; Sareen, S. Breeding for heat tolerance in wheat. In *Recent Trends on Production Strategies of Wheat in India*; Shukla, R.S., Mishra, P.C., Chatrath, R., Gupta, R.K., Tomar, S.S., Sharma, I., Eds.; JNKVV, Jabalpur & ICAR-IIWBR: Karnal, Indi, 2014; pp. 15–29.

33. Ni, A.; Li, H.; Zhao, Y.; Peng, H.; Hu, Z.; Xin, M.; Sun, Q. Genetic improvement of heat tolerance in wheat: Recent progress in understanding the underlying molecular mechanisms. *Crop J.* **2017**, *6*, 32–41. [CrossRef]

34. Kabbaj, H.; Sall, A.T.; Al-Abdallat, A.; Geleta, M.; Amri, A.; Filali-Maltouf, A.; Bassi, F.M. Genetic diversity within a global panel of durum wheat (*Triticum durum* L.) landraces and modern germplasm reveals the history of alleles exchange. *Front. Plant Sci.* **2017**, *8*, 1277. [CrossRef] [PubMed]

35. Sall, A.T.; Bassi, F.M.; Cisse, M.; Gueye, H.; Ndoye, I.; Filali-Maltouf, A.; Ortiz, R. Durum wheat breeding: In the heat of the Senegal river. *Agriculture* **2018**, *8*, 99. [CrossRef]

36. Bates, D.; Maechler, M.; Bolker, B.; Walker, S. Fitting linear mixed-effects models using lme4. *J. Stat. Softw.* **2015**, *67*, 1–48. [CrossRef]

37. R Development Core Team. *R: A Language and Environment for Statistical Computing*; R Foundation For Statistical Computing: Vienna, Austria, 2016.

38. DeLacy, I.H.; Basford, K.E.; Cooper, M.; Bull, J.K.; McLaren, C.G. Analysis of multi-environment trials: An historical perspective. In *Plant Adaptation and Crop Improvement*; Cooper, M., Hammer, G.L., Eds.; CAB International: Oxfordshire, UK, 1996.

39. Wickham, H. *ggplot2: Elegant Graphics for Data Analysis*; Springer: New York, NY, USA, 2009.

40. Wei, T.; Simko, V. *R package "corrplot": Visualization of a Correlation Matrix (Version 0.84)*; R Foundation for Statistical Computing: Vienna, Austria, 2017.

41. Wood, S.N. *Generalized Additive Models: An Introduction with R*, 2nd ed.; Chapman & Hall, CRC: Boca Raton, Florida, 2017; ISBN 9781498728331.

42. Wood, S.N.; Pya, N.; Saefken, B. Smoothing parameter and model selection for general smooth models. *J. Am. Stat. Assoc.* **2016**, *111*, 1548–1575. [CrossRef]

43. Fisher, R.A.; Maurer, R. Drought resistance in spring wheat cultivars I. grain yield responses. *Aust. J. Agric. Res.* **1978**, *29*, 897–917. [CrossRef]

44. Nachit, M.M.; Ouassou, A. Association of yield potential, drought tolerance and stability of yield in *Triticum turgidum* var *durum*. In Proceedings of the 7th International Wheat Symposium, Cambridge, UK, 13–19 July 1988; pp. 867–870.

45. Rosielle, A.A.; Hamblin, J. Theoretical aspects of selection for yield in stress and non-stress environments. *Crop Sci.* **1981**, *21*, 943–946. [CrossRef]

46. Bivand, R.; Ono, H.; Dunlap, R.; Stigler, M. R Package Classint: Choose Univariate Class Intervals. 2018. Available online: http://cran.r-project.org/package=classInt (accessed on 10 September 2019).

47. Lenth, R.; Love, J.; Maxime, H. Estimated Marginal Means, aka Least-Squares Means. 2017. Available online: http://cran.r-project.org/package=emmeans (accessed on 14 June 2019).

48. Maccaferri, M.; Harris, N.S.; Twardziok, S.O.; Pasam, R.K.; Gundlach, H.; Spannagl, M.; Himmelbach, A. Durum wheat genome highlights past domestication signatures and future improvement targets. *Nat. Genet.* **2019**, *51*, 885–895. [CrossRef] [PubMed]

49. Bradbury, P.J.; Zhang, Z.; Kroon, D.E.; Casstevens, T.M.; Ramdoss, Y.; Buckler, E.S. TASSEL: Software for association mapping of complex traits in diverse samples. *Bioinformatics* **2007**, *23*, 2633–2635. [CrossRef]

50. Jujumaan. Estimating and Plotting the Decay of Linkage Disequilibrium. Available online: https://jujumaan.com/2017/07/15/linkage-disequilibrium-decay-plot/ (accessed on 15 July 2017).

51. Bassi, F.; Brahmi, H.; Sabraoui, A.; Amri, A.; Nsarellah, N.; Nachit, M.M.; Al-Abdallat, A.; Chen, M.S.; Lazraq, A.; El Bouhssini, M. Genetic identification of loci for Hessian fly resistance in durum wheat. *Mol. Breed.* **2019**, *39*, 24. [CrossRef]

52. Sukumaran, S.; Xiang, W.; Bean, S.R.; Pedersen, J.F.; Kresovich, S.; Tuinstra, M.R.; Yu, J. Association mapping for grain quality in a diverse sorghum collection. *Plant Genom. J.* **2012**, *5*, 126–135. [CrossRef]

53. Duggal, P.; Gillanders, E.M.; Holmes, T.N.; Bailey-Wilson, J.E. Establishing an adjusted p-value threshold to control the family-wide type 1 error in genome wide association studies. *BMC Genom.* **2008**, *9*, 516. [CrossRef] [PubMed]

54. Pearson, K. Notes on regression and inheritance in the case of two parents. *Proc. R. Soc. Lond.* **1985**, *58*, 240–242.

55. Sofield, I.; Evans, L.T.; Cook, M.G.; Wardlaw, I.F. Factors influencing the rate and duration of grain filling in wheat. *Aust. J. Plant Physiol.* **1977**, *4*, 785–797. [CrossRef]

56. Gooding, M.J.; Ellis, R.H.; Shewry, P.R.; Schofield, J.D. Effects of restricted water availability and increased temperature on the grain filling, drying and quality of winter wheat. *J. Cereal Sci.* **2003**, *37*, 295–309. [CrossRef]

57. Corbellini, M.; Canevara, M.G.; Mazza, L.; Ciaffi, M.; Lafiandra, D.; Tozzi, L.; Borghi, B. Effect of the duration and intensity of heat shock during grain filling on dry matter and protein accumulation, technological quality and protein composition in bread and durum wheat. *Aust. J. Plant Physiol.* **1997**, *24*, 245–250. [CrossRef]

58. Corbellini, M.; Mazza, L.; Ciaffi, M.; Lafiandra, D.; Borghi, B. Effect of heat shock during grain filling on protein composition and technological quality of wheats. *Euphytica* **1998**, *100*, 147–154. [CrossRef]

59. Ugarte, C.; Calderini, D.F.; Slafer, G.A. Grain weight and grain number responsiveness to pre-anthesis temperature in wheat, barley and triticale. *Field Crops Res.* **2006**, *100*, 240–248. [CrossRef]

60. Nachit, M.M. Durum breeding for Mediterranean drylands of North Africa and West Asia. In *Durum Wheat: Challenges and Opportunities*; Rajaram, S., Saari, E., Hettel, G.P., Eds.; Wheat Special Report; CIMMYT: Ciudad Obregon, Mexico, 1992; Volume 9.

61. Barnabas, B.; Jager, K.; Feher, A. The effect of drought and heat stress on reproductive processes in cereals. *Plant Cell Environ.* **2008**, *31*, 11–38. [CrossRef]

62. Reynolds, M.; Foulkes, J.; Furbank, R.; Griffiths, S.; King, J.; Murchie, E.; Parry, M.; Slafer, G. Achieving yield gains in wheat. *Plant Cell Environ.* **2012**, *35*, 1799–1823. [CrossRef]

63. Masoni, A.; Ercoli, L.; Mariotti, M.; Arduini, I. Post-anthesis accumulation and remobilization of dry matter, nitrogen and phosphorus in durum wheat as affected by soil type. *Eur. J. Agron.* **2007**, *26*, 179–186. [CrossRef]

64. Prasad, P.V.V.; Pisipati, S.R.; Mutava, R.N.; Tuinstra, M.R. Sensitivity of grain sorghum to high temperature stress during reproductive development. *Crop Sci.* **2008**, *48*, 1911–1917. [CrossRef]

65. Sukumaran, S.; Lopes, M.; Dreisigacker, S.; Reynolds, M. Genetic analysis of multi-environmental spring wheat trials identifies genomic regions for locus-specific trade-offs for grain weight and grain number. *Theor. Appl. Genet* **2017**, *131*, 1–14. [CrossRef]

66. Sio-Se Mardeh, A.; Ahmadi, A.; Poustini, K.; Mohammadi, V. Evaluation of drought resistance indices under various environmental conditions. *Field Crops Res.* **2006**, *98*, 222–229. [CrossRef]

67. Dodig, D.; Zoric, M.; Kandic, V.; Perovic, D.; Surlan-Momirovic, G. Comparison of responses to drought stress of 100 wheat accessions and landraces to identify opportunities for improving wheat drought resistance. *Plant Breed.* **2012**, *131*, 369–379. [CrossRef]

68. Mason, R.E.; Mondal, S.; Beecher, F.W.; Pacheco, A.; Jampala, B.; Ibrahim, A.M.H.; Hays, D.B. QTL associated with heat susceptibility index in wheat (*Triticum aestivum* L.) under short-term reproductive stage heat stress. *Euphytica* **2010**, *174*, 423–436. [CrossRef]

69. Mason, R.E.; Mondal, S.; Beecher, F.W.; Hays, D.B. Genetic loci linking improved heat tolerance in wheat (*Triticum aestivum* L.) to lower leaf and spike temperature under controlled conditions. *Euphytica* **2011**, *180*, 181–194. [CrossRef]

70. Orcen, N.; Altinbas, M. Use of some stress tolerance indices for late drought in spring wheat. *Fresenius Environ. Bull.* **2014**, *24*.

71. Clarke, J.M.; Depauw, R.M.; Townley-Smith, T.F. Evaluation of methods for quantification of drought tolerance in wheat. *Crop Sci.* **1992**, *32*, 723–728. [CrossRef]

72. Rekika, D.; Kara, Y.; Souyris, I.; Nachit, M.M.; Asbati, A.; Monneveux, P. The tolerance of PSII to high temperatures in durum wheat (*T. Turgidum conv. Durum*): Genetic variation and relationship with yield under heat stress. *Cereal Res. Commun.* **2000**, *28*, 395–402.

73. Lopes, M.S.; Reynolds, M.P.; Jalal-Kamali, M.R.; Moussa, M.; Feltaous, Y.; Tahir, I.S.A.; Barma, N.; Vargas, M.; Mannes, Y.; Baum, M. The yield correlations of selectable physiological traits in a population of advanced spring wheat lines grown in warm and drought environments. *Field Crops. Res.* **2012**, *128*, 129–136. [CrossRef]

74. Reynolds, M.; Tuberosa, R. Translational research impacting on crop productivity in drought-prone environments. *Curr. Opin. Plant Biol.* **2008**, *11*, 171–179. [CrossRef] [PubMed]

75. Heidari, B.; Sayed-Tabatabaei, B.E.; Saeidi, G.; Kearsey, M.; Suenaga, K. Mapping QTL for grain yield, yield components, and spike features in a doubled haploid population of bread wheat. *Genome* **2011**, *54*, 517–527. [CrossRef] [PubMed]

76. Tahmasebi, S.; Heidari, B.; Pakniyat, H.; McIntyre, C.L. Mapping QTLs associated with agronomic and physiological traits under terminal drought and heat stress conditions in wheat (*Triticum aestivum* L.). *Genome* **2017**, *60*, 26–45. [CrossRef] [PubMed]

77. Quarrie, S.A.; Steed, A.; Calestani, C.; Semikhodskii, A.; Lebreton, C.; Chinoy, C.; Schondelmaier, J. A high-density genetic map of hexaploid wheat (*Triticum aestivum* L.) from the cross Chinese Spring x SQ1 and its use to compare QTL for grain yield across a range of environments. *Theor. Appl. Genet.* **2005**, *110*, 865–880. [CrossRef] [PubMed]

78. Acuna-Galindo, M.A.; Mason, R.E.; Subrahmanyam, N.K.; Hays, D. Meta-analysis of wheat QTL regions associated with adaptation to drought and heat stress. *Crop Sci.* **2014**, *55*, 477–492. [CrossRef]

79. McIntyre, C.L.; Mathews, K.L.; Rattey, A.; Drenth, J.; Ghaderi, M.; Reynolds, M.; Chapman, C.S.; Shorter, R. Molecular detection of genomic regions associated with grain yield and yield components in an elite bread wheat cross evaluated under irrigated and rainfed conditions. *Theor. Appl. Genet.* **2010**, *120*, 527–541. [CrossRef] [PubMed]

80. Araus, J.L.; Bort, J.; Steduto, P.; Villegas, D.; Royo, C. Breeding cereals for Mediterranean conditions: Ecophysiological clues for biotechnology application. *Ann. Appl. Biol.* **2003**, *142*, 129–141. [CrossRef]

81. Slafer, G.A.; Araus, J.L.; Royo, C.; García Del Moral, L.F. Promising eco-physiological traits for genetic improvement of cereal yields in Mediterranean environments. *Ann. Appl. Biol.* **2005**, *146*, 61–70. [CrossRef]

82. McCartney, C.A.; Somers, D.J.; Humphreys, D.G.; Lukow, O.; Ames, N.; Noll, J.; Cloutier, S.; McCallum, B.D. Mapping quantitative trait loci controlling agronomic traits in the spring wheat cross RL4452 9 'AC Domain'. *Genome* **2005**, *48*, 870–883. [CrossRef]

83. Collins, N.C.; Tardieu, F.; Tuberosa, R. Quantitative trait loci and crop performance under abiotic stress: Where do we stand? *Plant Physiol.* **2008**, *147*, 469–486. [CrossRef]

84. Rebetzke, G.J.; van Herwaarden, A.F.; Jenkins, C.; Weiss, M.; Lewis, D.; Ruuska, S.; Tabe, L.; Fettell, N.A.; Richards, R.A. Quantitative trait loci for water soluble carbohydrates and associations with agronomic traits in wheat. *Aust. J. Agric. Res.* **2008**, *59*, 891–905. [CrossRef]

85. Zaim, M.; El Hassouni, K.; Gamba, F.; Filali-Maltouf, A.; Belkadi, B.; Ayed, S.; Amri, A.; Nachit, M.; Taghouti, M.; Bassi, F. Wide crosses of durum wheat (*Triticum durum* Desf.) reveal good disease resistance, yield stability, and industrial quality across Mediterranean sites. *Field Crops Res.* **2017**, *214*, 219–227. [CrossRef]

86. Bassi, F.; Sanchez-Garcia, M. Adaptation and stability analysis of ICARDA durum wheat elites across 18 countries. *Crop Sci.* **2017**, *57*, 2419–2430. [CrossRef]

87. Graziani, M.; Maccaferri, M.; Royo, C.; Salvatorelli, F.; Tuberosa, R. QTL dissection of yield components and morpho-physiological traits in a durum wheat elite population tested in contrasting thermo-pluviometric conditions. *Crop Pasture Sci.* **2014**, *65*, 80–95. [CrossRef]

Article

Screening of Diverse Ethiopian Durum Wheat Accessions for Aluminum Tolerance

Edossa Fikiru Wayima [1], Ayalew Ligaba-Osena [2], Kifle Dagne [3], Kassahun Tesfaye [3], Eunice Magoma Machuka [4], Samuel Kilonzo Mutiga [4] and Emmanuel Delhaize [5,*]

[1] Department of Plant Sciences, College of Agriculture and Natural Resources, Madda Walabu University, Bale-Robe P.O. Box 247, Ethiopia
[2] Laboratory of Molecular Biology and Biotechnology, The University of North Carolina at Greensboro, 321 McIver St., 308 Sullivan Science Building, Greensboro, NC 27412, USA
[3] Cellular and Molecular Biology Department, College of Natural Sciences, Addis Ababa University, Addis Ababa P.O. Box 1176, Ethiopia
[4] Biosciences Eastern and Central Africa-International Livestock Research Institute (BecA-ILRI) Hub, Nairobi P.O. Box 30709-00100, Kenya
[5] CSIRO Agriculture and Food, GPO Box 1700, Canberra 2601, Australia
* Correspondence: manny.delhaize@csiro.au

Received: 15 July 2019; Accepted: 7 August 2019; Published: 9 August 2019

Abstract: Acid soils and associated Al^{3+} toxicity are prevalent in Ethiopia where normally Al^{3+}-sensitive durum wheat (*Triticum turgidum* ssp *durum* Desf.) is an important crop. To identify a source of Al^{3+} tolerance, we screened diverse Ethiopian durum germplasm. As a center of diversity for durum wheat coupled with the strong selection pressure imposed by extensive acid soils, it was conceivable that Al^{3+} tolerance had evolved in Ethiopian germplasm. We used a rapid method on seedlings to rate Al^{3+} tolerance according to the length of seminal roots. From 595 accessions screened using the rapid method, we identified 21 tolerant, 180 intermediate, and 394 sensitive accessions. When assessed in the field the accessions had tolerance rankings consistent with the rapid screen. However, a molecular marker specific for the D-genome showed that all accessions rated as Al^{3+}-tolerant or of intermediate tolerance were hexaploid wheat (*Triticum aestivum* L.) that had contaminated the durum grain stocks. The absence of Al^{3+} tolerance in durum has implications for how Al^{3+} tolerance evolved in bread wheat. There remains a need for a source of Al^{3+}-tolerance genes for durum wheat and previous work that introgressed genes from bread wheat into durum wheat is discussed as a potential source for enhancing the Al^{3+} tolerance of durum germplasm.

Keywords: aluminum; resistance; Ethiopia; durum; hydroponics; soil acidity; *Triticum turgidum* ssp. *durum* Desf.

1. Introduction

Durum (*Triticum turgidum* ssp *durum* Desf.) wheat and bread wheat (*Triticum aestivum* L.) provide the Ethiopian population with a large proportion of its caloric intake [1,2]. Despite a general increase in production and productivity of wheat during the last two decades (1998–2016), Ethiopia imports a substantial amount of both durum and bread wheat. Durum production as a proportion of the total wheat produced in Ethiopia has declined over the past few decades from about 80% in the 1980s to an estimated current proportion of only about 20%. Despite a reduction in the proportion of grain produced in Ethiopia, durum constitutes 50% to 80% of the wheat grain imported in any given year [1], indicating that demand for durum remains strong. One of the abiotic stresses that may be contributing to the decline in durum production is the prevalence of acid soils in Ethiopia. Durum wheat is very sensitive of the toxic Al^{3+} found in acid soil and is the most sensitive of the small-grained crops to Al^{3+}

toxicity [3]. By contrast, bread wheat has Al^{3+}-tolerant alleles of *TaALMT1* [4] and *TaMATE1B* [5] along with other yet to be cloned Al^{3+}-tolerance genes [6].

Worldwide, acidic soils are one of the most important limitations to agricultural production [7]. In Ethiopia, acid soils account for about 34% of agricultural land area that range from slight to strongly acidic soils [8]. Acid soils generally occupy the western part of the country extending from southwest to northwest, although strongly acidic soils occur mainly in the western part of the country including the lowlands. Acid soils are particularly prevalent in the highlands of Ethiopia [9] and the application of lime commonly results in improved yields of various crops [10,11]. The productivity of wheat over acidic areas of Ethiopia is low as compared to parts of the country where soil acidity is absent. For instance, in areas with strongly acidic soils, which occur widely in the western and southwestern parts of the country, the productivity of wheat is as low as 0.8–2.0 t/ha. These areas include West Wallaga, Illu-Ababora, Jimma, Gamo Gofa, Asosa, and Metekel zones. On the other hand, a relatively higher grain yield, ranging from 2.5 to 3.1 t/ha, is obtained in regions with near neutral soil pH including West Arsi, Arsi, Siltie, East Shewa, and Bale zones (Supplementary Materials file, Table S1).

Aluminum is solubilized in acid soils into the toxic Al^{3+} which in sensitive plants typically affects the viability of the root apex. Al^{3+} inhibits root growth resulting in reduced uptake of water and nutrients ultimately hindering plant growth and development [12]. In addition to its toxic effects on plant tissues, Al^{3+} affects nutrient availability within the soil. All of these effects significantly reduce crop yield. While management, primarily by application of lime, is important for neutralizing acid soils, this can be costly and it can take years to correct acidity at depth. A complementary strategy for improving crop production on acid soils is the use of Al^{3+}-tolerant germplasm developed through breeding or genetic modification along with liming practices. In several crop species, variation in Al^{3+} tolerance has been identified and selective breeding programs can be implemented to increase production on acid soils [13–15]. However as noted above, durum wheat is one of the most Al^{3+} sensitive of the small-grained crops and shows little variation in Al^{3+} tolerance [16]. Bread wheat is hexaploid and possesses the A-, B-, and D-genomes whereas durum wheat is tetraploid possessing only the A- and B-genomes. Although, as noted below, a major gene for Al^{3+} tolerance is found on the D-genome of bread wheat, there are other genes for Al^{3+} tolerance in bread wheat found on the A- and B-genomes [6]. Early reports that identified Al^{3+}-tolerant durum genotypes with a level of tolerance similar to that of bread wheat [17] can now be attributed to misidentification or contamination of durum grain stocks with bread wheat [18], highlighting the importance of verifying the genetic identity of germplasm. Bread wheat, in contrast to durum, shows a large variation in Al^{3+} tolerance and much of this is conditioned by alleles of the *TaALMT1* gene located on chromosome 4D [19] and *TaMATE1B* located on chromosome 4B. *TaALMT1* encodes an Al^{3+}-activated membrane channel permeable to malate with the malate exuded by root apices binding toxic Al^{3+} to protect the developing roots. Another Al^{3+} tolerance gene in bread wheat is *TaMATE1B* that encodes a citrate transporter in the plasma membrane and in some bread wheat genotypes confers a lower level of tolerance than *TaALMT1* [20] whereas in other genotypes *TaMATE1B* appears to be the predominant Al^{3+} tolerance gene [21]. Both *TaALMT1* and *TaMATE1B* have been introgressed from bread wheat into durum wheat. This required use of the pairing homeologous (*ph1c*) mutant to enable *TaALMT1* to be transferred from chromosome 4D in bread wheat to chromosome 4B in durum wheat [20]. In contrast to bread wheat, for durum seedlings grown on acid soil, *TaMATE1B* appears to provide a greater level of Al^{3+} tolerance than *TaALMT1* for reasons that are not understood.

Ethiopia has been considered as a center of diversity for durum germplasm and landraces grown by farmers are a potential source of agronomically important genes [22,23]. A 90 K single nucleotide polymorphism (SNP) chip analysis supports the notion that durum wheat in Ethiopia is particularly diverse [24]. Furthermore, others undertaking similar phylogenetic analyses based on SNP markers have suggested that Ethiopia is a second center of origin for durum wheat [25]. We speculated that because of the diversity of the durum germplasm coupled with a strong selection pressure imposed by extensive areas of acid soils in Ethiopia useful levels of Al^{3+} tolerance could have evolved in Ethiopian durum landraces. Cultivated durum genotypes sourced from various countries were found to be all Al^{3+} sensitive [16] whereas others [26] identified relatively tolerant lines in a similar population. The Ethiopian landraces are likely to be more diverse in their genetic makeup than the populations previously screened for tolerance and provide an opportunity to identify unique genes that confer tolerance of acid soils and Al^{3+} toxicity in particular. As an initial strategy it would be preferable to identify Al^{3+} tolerance in germplasm already adapted to Ethiopian conditions instead of introgressing genes from other sources. In this study we screened a diverse set of Ethiopian durum germplasm for Al^{3+} tolerance using a rapid hydroponic screen and then assessed selected lines more thoroughly in the hydroponic screen and finally in field trials on acid soil. Despite Ethiopia having conditions that would favor the evolution of Al^{3+} tolerance, the diverse durum germplasm was found to be Al^{3+} sensitive and highlighted the importance of verifying the identity of grain stocks. The absence of Al^{3+} tolerance genes in diverse durum germplasm suggests that genes encoding Al^{3+} tolerance found on the A- and B-genomes of bread wheat arose subsequent to the hybridization with the D-genome that produced hexaploid wheat.

2. Materials and Methods

2.1. Germplasm

A total of 595 durum wheat accessions obtained from the Ethiopian Biodiversity Institute (EBI: www.ibc.gov.et; Supplementary Materials file, Table S2), were screened in a series of non-replicated trials for their Al^{3+} tolerance using a rapid hydroponic screen. Based on their performance at this preliminary stage, accessions were classified as Al^{3+}-tolerant, -intermediate, or -sensitive (see below). Subsequently all the tolerant as well as selected accessions of the intermediate and sensitive classes, totaling 150 accessions, were evaluated in a replicated hydroponic experiment and a field trial. For the experiments that screened previously identified Al^{3+}-tolerant lines of durum the selected lines are described by Raman et al. [26]. A durum line (Langdon 4D (4B)) that has the 4B chromosome substituted by the 4D chromosome of bread wheat [27] was included as a positive control of a confirmed Al^{3+}-tolerant durum line [18].

2.2. Hydroponic Culture

A rapid hydroponic screening was undertaken using an apparatus comprised of a plastic basin (for holding nutrient solution), a plate that held seed of different wheat accessions separately submerged in nutrient solution and an aeration system (Supplementary Materials file, Figure S1). The method is based on one described previously [16]. Seedlings grown submerged in the aerated nutrient solution remained viable and could be transplanted to soil when required to bulk up grain. Grain harvested from single plants of selected seedlings from this "preliminary" screen was then used in a replicated hydroponic experiment using the same growth conditions. For the replicated experiment, the average performance of 10 seedlings for each accession was determined and the experiment repeated three times.

The nutrient solution comprised of 500 µM KNO_3, 500 µM $CaCl_2$, 500 µM NH_4NO_3, 150 µM $MgSO_4$, 10 µM KH_2PO_4, 2 µM $FeCl_3$, and 5 µM of $Al_2SO_4.18H_2O$ as described previously [28]. The pH of the solution was adjusted to about 4.3 with 1 M HCl. Dry grain was immersed in the solution and seedlings allowed to grow submerged for 5 days in the solution culture with the nutrient solution changed every day to maintain pH and Al^{3+} concentration relatively constant. After 5 days seedlings had typically developed three seminal roots and the total root length was recorded to the nearest 0.1 cm.

We used conventional hydroponics with the same nutrient solution composition as described above to compare the Al^{3+} tolerance of a set of durum and bread wheat genotypes in a range of $AlCl_3$ concentrations. Relative root length for the various genotypes was calculated after 3 days growth as root length in solution that contained Al^{3+} relative to root length in control solution that lacked Al^{3+}. Errors associated with relative root length were calculated as described previously [29].

2.3. Determining the Identity of the Al³⁺-Tolerant Accessions

The morphological similarity between durum and bread wheat made it difficult to establish whether the identified Al^{3+}-tolerant lines were durum or bread wheat by using phenotypic traits alone. Therefore, a polymerase chain reaction (PCR) assay that targeted the Dgas44 sequence was used to determine whether the lines were durum or bread wheat [30]. Dgas44 is a D-genome specific repetitive sequence that can be used to distinguish hexaploid wheat from tetraploid wheats that lack the D-genome [31]. The sequence of forward and reverse primers of Dgas44 marker, respectively, were 5′-CTTCTGACGGGTCAGGGGCAC-3′ and 5′-CTGAATGCCCCTGCGGCTTAAG-3′.

Ten grain of bulked up samples used in the field trial along with one verified bread wheat cultivar (Enkoy) and one durum wheat accession (8317), distinguished by its reddish/pinkish seed color, were planted in pots. Enkoy variety was included as a positive control (i.e., possesses the D-genome since it is a bread wheat cultivar), while accession 8317 was used as a negative control since it is a known durum accession that lacks the D-genome. Young green leaves were collected separately from three individual plants for DNA extraction. Leaf samples were freeze-dried in liquid nitrogen and pulverized with a Geno/Grinder 2000 and genomic DNA was extracted with a ZymoResearch kit (Plant/Seed MiniPrep) following the manufacturer's protocol. The extracted genomic DNA was quantified with a NanoDrop 2000 UV-Vis Spectrophotometer (Thermo Scientific, Waltham, MA, USA) and PCR of samples undertaken as previously described [18].

2.4. Field Experiment

The field experiment was conducted at Bedi (38°36′3″ E, 9°5′59″ N), which is located in Watabicha Minjaro Kebele, Welmera District, West Shewa Zone, Oromia Regional State, Ethiopia (Supplementary Materials file, Figure S2). It is situated at approximately 35 km west of Addis Ababa and about 25 km away from Holeta town in the north-east. Crops are generally planted from around mid-June, though it varies with crop type and the time of onset of rains.

Samples of the reddish-brown soil to a depth of 20 cm were collected from the experimental plot with an auger in a regular pattern following a line transect. The soil samples were submitted the following day to JIJE LABOGLASS PLC (Addis Ababa, Ethiopia) for analysis of the major parameters using standard procedures as shown in Table 1. Characterized as a clay type, the pH of a water extract of the soils was found to be 4.92 indicating acidity and the likely presence of Al^{3+} toxicity. Note that a water extract will generally have a higher pH value than if a $CaCl_2$ solution is used for the extraction. For the limed treatments 451 kg of $CaCO_3$ was manually applied to the soil the year prior to the trial covering an area of 200 m². The lime was obtained from Guder Lime Factory (Guder, West Shewa Zone, Ethiopia) through collaboration with Oromia Agriculture and Rural Development Bureau. The plot was then ploughed immediately and re-ploughed after a week to thoroughly mix the lime with the soil to a depth of about 20 cm so that Al^{3+} would be detoxified in the soil solution.

A total of 150 accessions were selected based on their response to Al^{3+} toxicity in the preliminary hydroponic screen described above. These accessions were planted during the major rainy season on limed and un-limed plots in the field using a randomized complete block design. The two treatments of the experiment (limed and un-limed blocks) were each replicated twice, and the replicated blocks were spaced 1 m apart such that the long edge of the blocks were arranged adjacent to one another. All 150 accessions were planted side by side in each block resulting in a total area of 30 × 2 m for each block. Individual accessions were planted within the blocks randomly in single rows 2 m long with 20 cm spacing between rows. Contamination of un-limed plots with lime through erosion and splashing was avoided by considering slope and spacing factors in the experimental layout. Specifically, plots at a higher slope were assigned as un-limed plots and were separated from the limed plots by 2 m. N and P fertilizers were applied after 3–4 weeks of planting as urea and di-ammonium phosphate at blanket recommendation rate (100 kg/ha) since there was no pre-determined site-specific fertilizer application rate. Hand-weeding was done at early seedling stage (about one month) and before the booting stage. Mature plants were harvested to measure total biomass and grain yield.

Table 1. Chemical and physical characteristics of the soils of the study area.

Soil Parameter		Value
pH (water extract; 1:1.25)		4.92
Buffer pH (water extract 1:2)		5.57
Electrical conductivity (dS/m)		0.07
Organic matter (%)		4.01
Total N (%)		0.78
Available P (mg/kg)		8.87
Exchangeable acidity (meq/100 g)		2.23
CEC and exchangeable bases (cmol (+)/kg)	CEC	27.19
	Ca	5.27
	Mg	0.66
	K	0.91
Texture	Clay (%)	59
	Silt (%)	28
	Sand (%)	13
	Soil class	Clay

2.5. Statistical Analysis

Normality of distributions of the replicated hydroponic and field data were tested with SPSS version 24 (IBM SPSS statistics for Windows 2016, Version 24.0. Armonk, NY, USA: IBM Corp.) and RStudio (version 1.0.143: Integrated Development for R. RStudio, Inc., Boston, MA, USA. Retrieved from www.rstudio.com). The non-parametric (Kruskal–Wallis) and parametric ANOVA were computed either with RStudio or SPSS in order to determine whether there was significant difference between accessions for their performance in Al^{3+} treatments. The general linear model in which both accessions and replications were considered as fixed factors was used for analysis of variance. The median absolute deviation, a non-parametric statistic, was computed with RStudio to assess the level of variability attributed to the trait. Furthermore, post-hoc analysis was conducted using Tukey's honestly significant difference for parametric statistics, while a nonparametric post-hoc was done with Dunn's test, a Kruskal–Wallis test based post-hoc with "agricolae" package of RStudio. The coefficient of determination (r^2) was calculated to examine the relationship between various variables.

3. Results

3.1. Hydroponic Screen

Durum wheat accessions (595) were initially screened for Al^{3+} tolerance using a rapid hydroponic method of submerged seedlings. These accessions sometimes comprised of two or more genotypes since there was a large variation in performance between individual plants of the accession (Figure 1). Furthermore, there were visually observable differences within an accession such as variation in grain color. To take this heterogeneity into account an accession was scored based on its best-performing seedling. The use of average performance of plants in representing an accession would have resulted in rejection of many accessions because of a poor average performance such that a single plant within the accession with an acceptable level of Al^{3+} tolerance would be lost.

Accessions were classified into three phenotypic classes based on their total root length as tolerant ($\geq$3.1 cm), intermediate ($\leq$3.0 cm but $\geq$2.1 cm) and susceptible ($\leq$2.0 cm). Moreover, accession 6956 was included in the tolerant class because of its exceptionally long primary seminal root. Using this criterion 21 accessions were classed as tolerant, 180 as intermediate, and 394 as susceptible to Al^{3+} toxicity (Supplementary Materials file, Table S3).

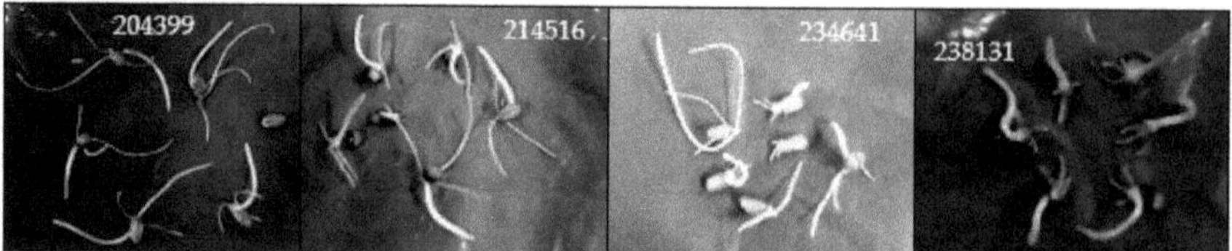

Figure 1. Examples of seedlings grown for 5 days in Al^{3+}-containing nutrient solution. Accessions 204399 and 214516 were scored as tolerant as was accession 234641 that was clearly segregating for tolerance. Accession 238131 was scored as sensitive. Individual seedlings identified within an accession to be tolerant were planted out to bulk up grain and the harvested grain used in a further hydroponic screen and a field trial.

A total of 150 accessions that included all 21 of the Al^{3+}-tolerant class, 79 of the best performers of the intermediate class, and a random selection of 50 of the sensitive class were evaluated in three replications to more precisely characterize their Al^{3+} tolerance. In this experiment accessions were represented in each experiment by the average performance of $\geq$10 seedlings using grain harvested from individually selected seedlings. The performance of accessions in the replicated experiment using the same hydroponic method as the preliminary screen was well correlated with the preliminary screen ($r^2 = 0.72$; Figure 2).

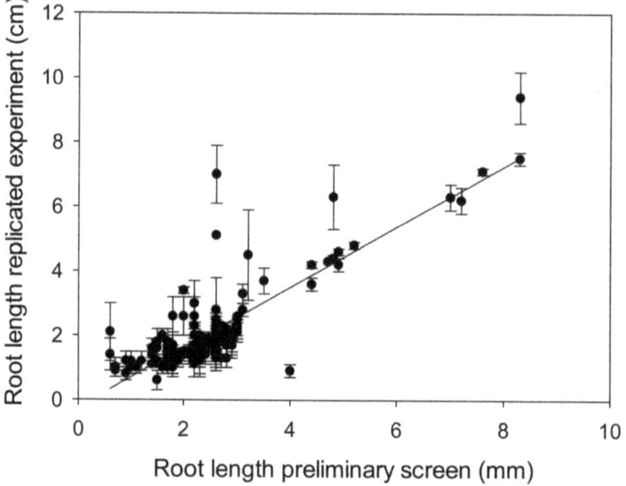

Figure 2. Relationship between total root length of the individual seedlings selected (total root length from the preliminary screen, $n = 1$ for each accession) that had the longest roots and the total root length of the resulting progeny of the selections showing the average of 10 or more seedlings (in three replications). Error bars indicate the standard error of the mean for the three replications.

3.2. Field Experiment

The set of 150 selections used in the replicated hydroponic screen was assessed in the field on an acid soil. As there was limited grain harvested from single plants of the selected seedlings, the accessions were assessed as single rows 2 m long on both an un-amended plot and an equivalent plot that had been amended with lime to neutralize the acidity. The grain yield and biomass means for the limed plots of all accessions combined was greater than the means of the acid plot indicating that soil acidity was clearly present at this site. Grain yield on the acid site was reduced on average to only 18% of the limed site whereas biomass was reduced to 28% of the limed site (Figure 3).

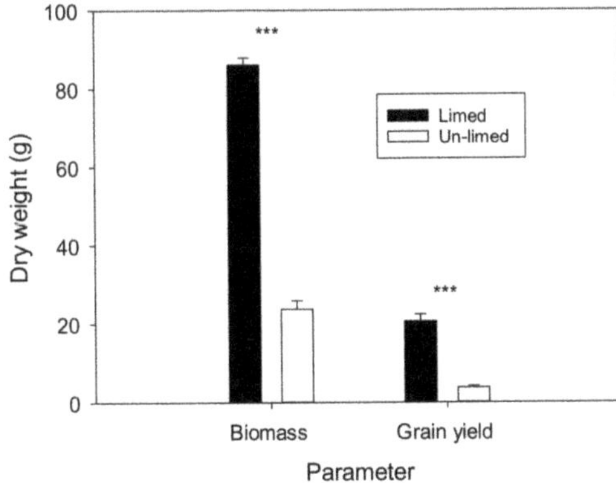

Figure 3. Liming improves biomass and grain yield of Ethiopian accessions grown on an acid soil. Ethiopian accessions (150) were grown in single rows 2 m long on an acid soil and the same soil that had lime incorporated to a depth of 20 cm. Final biomass and grain yield were determined and data of all accessions were combined with error bars indicating the standard error of the mean (n = 150). Student's *t*-test indicated significant differences between treatments (***; $p < 0.001$).

The performance of the Al^{3+}-tolerant lines on the un-limed acid plot was remarkable and clearly distinct from that of the sensitive accessions. Figure 4 shows the relative (un-limed/limed) biomass and grain yields of the accessions selected for the field trial against the root length as determined in the preliminary screen.

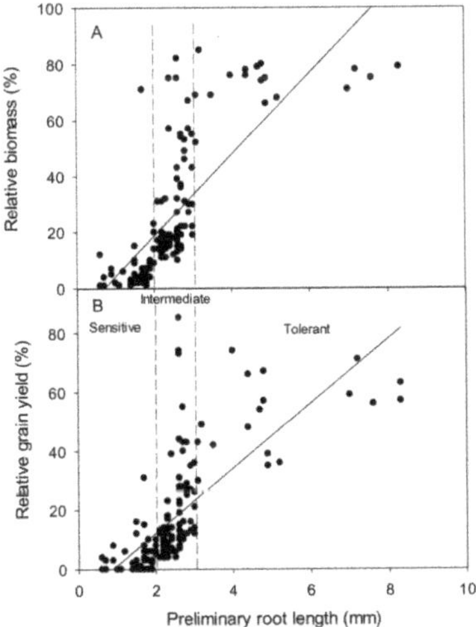

Figure 4. Root length as determined with the rapid screen identifies Al^{3+}-tolerant accessions based on both mature biomass (**A**) and grain yield (**B**). The preliminary root length is a measure of the total root length of the most tolerant seedling in the pool of seedlings assessed for each accession. The dashed

vertical lines denote the cut-off points for root lengths of seedlings initially classed as; tolerant: >3.0 cm; intermediate: 2.0–3.0 cm; sensitive: <2.0 cm. The single seedling was grown to seed and the resulting progeny grown in two replicates as 2 m rows on un-limed and limed plots. The mean of the replicates was used to calculate the relative biomass and relative grain yields where values for the un-limed plots were divided by the limed plots and multiplied by 100 to express data as a percentage. The solid line shows a linear regression with $r^2 = 0.56$ for (A) and $r^2 = 0.52$ for (B).

Expressing data in relative terms takes into account inherent differences in plant vigor when Al^{3+} is absent but a similar relationship was found when using unmodified data of only the acid plot (Supplementary Materials file, Figure S3). Biomass and final grain yield of genotypes grown in the acid trial were strongly correlated (Figure 5A). When biomass and grain yield data were combined within each grouping of lines selected from the preliminary screen (sensitive, intermediate, and tolerant), they showed rankings consistent with their grouping based on the screen (Figure 5B).

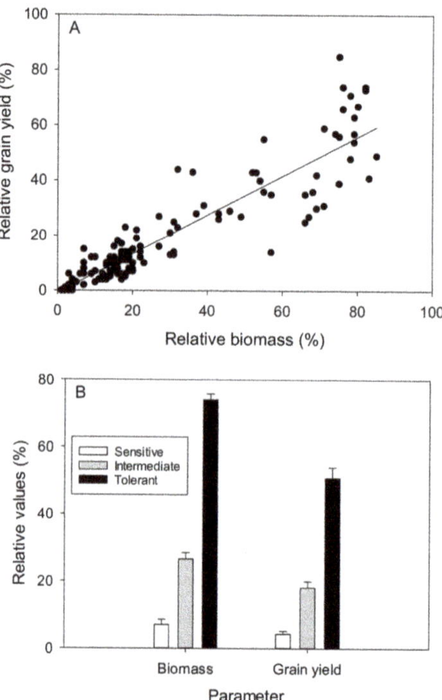

Figure 5. Relative biomass and relative grain yield are highly correlated and the classification of accessions into groups with the preliminary screen of Al^{3+}-tolerant, -intermediate, and -sensitive is consistent with final biomass and grain yield. Biomass and grain yield of accessions grown on the un-limed plot are expressed as a percent of the limed plot using the procedure described in the legend of Figure 4. (**A**) Relationship between relative biomass and relative grain yield for all accessions grown in the field ($r^2 = 0.84$). (**B**) Relative biomass and relative grain yield of all accessions allocated to the three classes were combined with error bars denoting the standard error and a one-way ANOVA of the data showed significant differences between all groups for each of biomass and grain yield (ANOVA on ranks $p < 0.001$ between classes; $n = 47$ for Sensitive, $n = 83$ for Intermediate, and $n = 20$ for Tolerant classes).

Since previous studies have found that stocks of durum wheat lines can be contaminated with bread wheat, we sought to verify that both Al^{3+}-tolerant and -intermediate selections were indeed durum wheat. It can sometimes be difficult to distinguish the species based solely on the phenotypes so we used the Dgas molecular marker that is unique to the D-genome to distinguish the species. Using Dgas we found that all the Al^{3+}-tolerant and -intermediate accessions were actually bread wheat (Figure 6), with only the most sensitive genotypes being durum wheat.

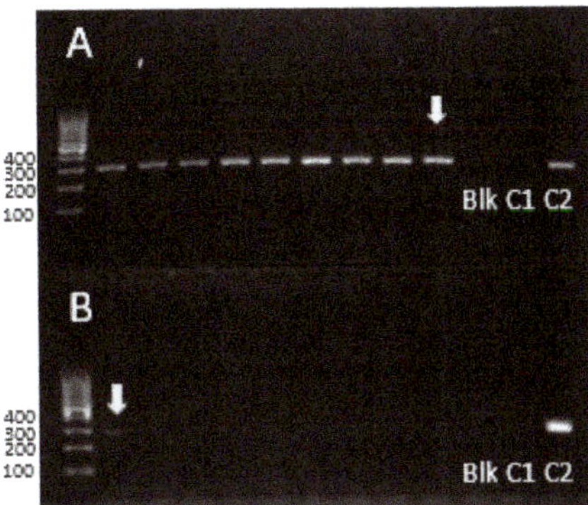

Figure 6. Examples of accessions screened by PCR for the Dgas sequence. Panel (**A**) is of Al^{3+}-tolerant selections while panel (**B**) are all Al^{3+}-sensitive selections except for the first lane after the DNA ladder. Most samples are shown as duplicates in lanes side by side except for the last lane of (A) (arrow) whose duplicate is the first lane of (B) (arrow). The lane on the left is a 100 bp DNA ladder with sizes of the smallest four markers shown on the right. A band at about 300 bp is indicative of the presence of the D-genome. Blk is a no DNA sample while C1 is a known durum wheat (accession 8317) and C2 is a known bread wheat variety (Enkoy). All lines (38 lines) that yielded more than 3.5 g per row in the field trial (Supplementary Materials file, Figure S3B) were analyzed and all possessed the 300 bp band indicating that they were hexaploid wheat.

3.3. Other Potential Sources of Al^{3+}-Tolerant Durum Wheat

As discussed in the Introduction, the Al^{3+} tolerance of durum has been enhanced by introgression of genes from bread wheat although not into Ethiopian germplasm [20]. As an alternate source of genes, a previous report identified three out of 420 tetraploid genotypes screened that were comparatively Al^{3+} tolerant although they did not approach the tolerance shown by a tolerant bread wheat used as a check line [26]. The lines had been confirmed as tetraploid with a molecular marker specific for the D-genome, so these genotypes were a potential source of genes that could be used in direct crosses to Ethiopian lines. To establish if these three lines had a useful level of Al^{3+} tolerance, we assessed their performance in hydroponic culture against bread wheat lines that varied in tolerance as well as a 4D (4B) substitution line of durum. However, all three lines were rated as sensitive with the best performer having a similar level of tolerance as Al^{3+}-sensitive bread wheat (Figure 7). By contrast, a durum line where the 4B chromosome was substituted by the 4D chromosome of hexaploid wheat had a level of Al^{3+} tolerance comparable to the tolerant hexaploids.

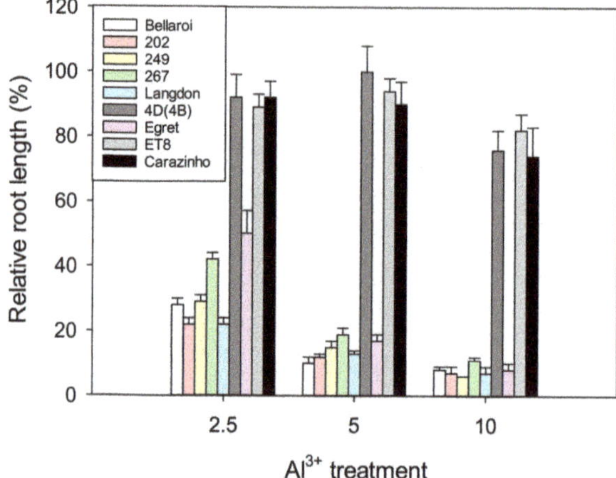

Figure 7. Relative Al^{3+} tolerance of durum wheat genotypes previously identified as being relatively Al^{3+} tolerant in comparison to a durum substitution line with chromosome 4D of bread wheat and bread wheat lines possessing Al^{3+}-tolerant and -sensitive alleles of the *TaALMT1* gene. Bellaroi is an Australian durum wheat cultivar; lines 202, 249, and 267 are accessions previously classified as being Al^{3+}-tolerant durum genotypes [26]; Langdon is the parental durum cultivar used to generate the 4D substitution line; 4D (4B) is a durum line in the Langdon background where chromosome 4B has been substituted with chromosome 4D; Egret is an Al^{3+}-sensitive bread wheat cultivar; ET8 and Carazinho are Al^{3+}-tolerant bread wheat lines. Root length is expressed as percent relative to a zero control and error bars indicate cumulative errors of three to five seedlings.

4. Discussion

4.1. A Rapid Screen Proves Robust and Correlates with Grain Yield in Field Trials when Ranking Wheat Germplasm for Al^{3+} Tolerance

In this work we screened Ethiopian durum accessions for Al^{3+} tolerance first using a high throughput method with hydroponics, then with a replicated hydroponic screen and finally in the field on acid soil. Ethiopian durum accessions that include landraces are reported to show great diversity for many traits and molecular analyses have verified the unique nature of this germplasm [24]. However, despite this diversity our work shows that there does not appear to be useful levels of Al^{3+} tolerance in Ethiopian durum wheat and that seedlings identified to be Al^{3+} tolerant were in fact contaminating bread wheat. It should be noted that many of the accessions classed as tolerant only had a few Al^{3+}-tolerant seedlings indicating that the durum grain stocks had been contaminated with bread wheat grain. This was verified by the use of a molecular marker that confirmed Al^{3+}-tolerant seedlings to be bread wheat. We conclude that the rapid screen was sufficiently robust as a preliminary screen to identify Al^{3+}-tolerant bread wheat contaminants. The accessions ranked as having intermediate tolerance in the preliminary screen could still have conferred a useful level of Al^{3+} tolerance for durum if they ranked similarly in the field. However, in the field the best performing accessions ranked as having intermediate Al^{3+} tolerance from the preliminary screen were all identified as bread wheat (Figure 6 and Supplementary Materials Figure S3B). Nevertheless, there exist Al^{3+}-sensitive genotypes of bread wheat so the rapid screen does not replace the use of a molecular marker in identifying bread wheat lines. Verifying the identity of durum wheat grain stocks is of particular importance for germplasm banks but should also be confirmed by researchers. Bread wheat genotypes in the past have been mistakenly identified as durum wheat and the most relevant to this study was the incorrect identification of Al^{3+}-tolerant durum lines that were subsequently found to be bread wheat [17,18].

In some cases, bread wheat has morphological characteristics in the field that are similar to durum wheat emphasizing the need for molecular analysis to establish the species identity [32].

The absence of Al^{3+} tolerance in unmodified *Triticum turgidum* ssp *durum* Desf. is consistent with previous studies that screened durum cultivars with presumably a lower level of diversity than the Ethiopian germplasm and also showed no or only a comparatively low level of Al^{3+} tolerance [16]. The simple and rapid screen undertaken with minimal equipment over only 5 days growth with seedlings submerged in nutrient solution showed a remarkable consistency in classifying the relative tolerance of germplasm grown in the field and measured for mature biomass and even grain yield (Figure 5B). Although there was a good general agreement of the hydroponic screens with performance of accessions in the field, there were exceptions. For instance, several accessions rated as having intermediate tolerance with the preliminary screen were the best or amongst the best for grain yield in the field (Figure 4B). This finding emphasizes the importance of verifying selections on acid soil whether in the field or in pots and that while hydroponic screens allow for rapid assessment of germplasm, the relative tolerance of genotypes can differ when grown in soil. For instance, a durum line carrying the *TaMATE1B* gene introgressed from hexaploid wheat (see below) shows marginal Al^{3+}-tolerance in hydroponics that is considerably less than a line carrying *TaALMT1*, whereas the situation is reversed when the lines are grown on acid soil, with *TaMATE1B* lines outperforming *TaALMT1* lines [20].

4.2. The Evolution of Al^{3+} Tolerance in Bread Wheat Occurred Subsequent to the Hybridization of the D-Genome

Here we show that durum lines previously reported to be relatively Al^{3+}-tolerant [26] had a level of tolerance similar to sensitive wheat (Figure 7), confirming the absence of useful Al^{3+} tolerance in durum wheat germplasm (excluding lines where genes from hexaploid wheat have been introgressed as discussed below). This observation is puzzling given reports of the presence of Al^{3+} tolerance genes on the A- and B-genomes of bread wheat. Although *TaALMT1*, the major gene for Al^{3+} tolerance in bread wheat, is located on chromosome 4D, there are several loci on the A- and B-genomes reported to confer Al^{3+} tolerance [6]. For example, *TaMATE1B* the only Al^{3+} tolerance gene other than *TaALMT1* that has been cloned from bread wheat, is located on chromosome 4B [33]. We surmise that the absence of Al^{3+} tolerance genes in diverse durum germplasm suggests that the multiple Al^{3+} tolerance genes found in bread wheat arose subsequent to the hybridization of the D-genome with the A- and B-genomes some 10,000 years ago. Transposable elements have been shown to enhance the level of expression of genes encoding transport proteins such as those of the MATE family [34] that confer Al^{3+} tolerance. The event of polyploidization commonly results in genome instability including the activation of transposable elements [35]. It is conceivable that activation of transposons when the hexaploid was formed has been key in enabling bread wheat to evolve Al^{3+} tolerance.

4.3. Strategies to Enhance the Al^{3+} Tolerance of Ethiopian Durum Wheat

To date the only durum germplasm verified to be Al^{3+} tolerant are those that were developed by introgression of Al^{3+} tolerance genes from bread wheat into durum wheat [20]. In those cases, the *ph1c* mutant was used to introgress the *TaALMT1* gene located on chromosome 4D of bread wheat and steps taken to avoid the hybrid necrosis that can occur when crossing bread to durum wheat to introgress the *TaMATE1B* gene on chromosome 4B. This germplasm is not considered to be genetically modified since it was developed by so-called natural means and can therefore be used in the field without restrictions. The germplasm is a source of Al^{3+} tolerance genes in a durum background that could be used to introgress one or both genes derived from bread wheat into selected Ethiopian durum germplasm. Many of the farms that grow durum wheat in Ethiopia are small holdings and it has been found that landraces otherwise known as "farmer's varieties" perform better than "improved" germplasm under many situations [36]. With this in mind it may be useful to cross the Al^{3+}-tolerant germplasm described above into selected landraces adapted to local regions. Using local landraces as recurrent parents in multiple backcrosses while tracking the presence of Al^{3+} tolerance genes using molecular markers should maintain the germplasm with various valuable landrace traits that could otherwise be lost [37]. The *TaMATE1B* gene seems a preferred candidate at this stage since it appears to be more effective than *TaALMT1* in a durum background and there exists a co-dominant marker that can be used to track the tolerant allele [20]. A recent publication has shown that the *TaMATE1B* gene introgressed into a durum genetic background confers a marked ability of roots from mature plants to withstand Al^{3+} toxicity when grown in an acid soil [38]. A simple backcrossing program where a single gene is introgressed into landraces could be useful in establishing whether or not *TaMATE1B* can improve durum grain production of landraces on acid soils of Ethiopia. We speculate that introgressing both *TaALMT1* with *TaMATE1B* may provide the greatest level of Al^{3+} tolerance to durum wheat as has been found for some bread wheat genotypes.

5. Conclusions

Here we show that despite the diversity of Ethiopian durum germplasm and large regions of acid soils, conditions conducive for the evolution of Al^{3+} tolerance, a useful level of Al^{3+} tolerance was not detected. A high throughput screen identified Al^{3+}-tolerant seedlings within accessions but all of them were shown to be contaminating bread wheat. This finding highlights the importance of correct identification of germplasm and confirmation of species identity with molecular markers particularly for closely-related species. The finding that Ethiopian durum germplasm is Al^{3+} sensitive is consistent with Al^{3+} tolerance having evolved in bread wheat subsequent to the hybridization of the D-genome with the A- and B-genomes. A field trial on acid soil showed biomass and grain yields that correlated with classifications based on a high throughput screen, confirming the utility of the rapid screen for preliminary assessment of germplasm.

Supplementary Materials: The following are available online at http://www.mdpi.com/2073-4395/9/8/440/s1, Figure S1: Photographs of the components of the hydroponic system. (A) The plastic cups to which dry grain was added. The holes allowed aeration of grain while the cups themselves separated the accessions from one another. The slots in the plastic stands at the end of a series of cups enabled the long air stone to be inserted and held firm. (B) Top view showing air stones and pumps along with the plastic basin used to hold the nutrient solution. (C) Top view of assembled equipment showing plastic cups with a partitioning plate inserted within each cup so that two accessions could be placed into each cup and separated from one another. Nutrient solution was added to the basin to fill and cover the cups that held the grain. The holes in the cups allowed sufficient aeration for growth of seedlings. Nutrient solution was changed every day to maintain pH at about 4.3 and Al_2SO_4 at 5 µM. Figure S2: Location of the study area in Ethiopia. Figure S3: Relationships between Al^{3+} tolerance determined with a rapid screen and (A) biomass or (B) grain yield for durum accessions grown on an acid plot in the field. Table S1: Wheat production across zones and special districts of Ethiopia showing yields averaged over 5 years. Table S2: Passport data of durum accessions used in hydroponics screening for Al^{3+} tolerance. Table S3: Preliminary screen of Al^{3+} tolerance of 594 Ethiopian durum wheat accessions grown in Al^{3+}-containing nutrient solution where total root length after 5 days growth was used to classify genotypes.

Author Contributions: Conceptualization, E.F.W.; methodology, E.F.W. and E.D.; formal analysis, E.F.W. and E.D.; investigation, E.F.W., E.D., A.L.O.; resources, E.F.W., A.L.O., K.T., E.M.M.; data curation, E.F.W.; writing—original draft preparation, E.F.W. and E.D.; writing—review and editing, A.L.O., and K.D.; supervision, E.D., A.L.O., K.D., K.T. and S.K.M.; project administration, K.T. and E.M.M.; funding acquisition, E.F.W., A.L.O. and K.T.

Funding: The hydroponic and field experiments were sponsored by Addis Ababa University and the Eastern Africa Agricultural Productivity Program, respectively, while the Dgas analysis was financed by the BecA-ILRI Hub through the Africa Biosciences Challenge Fund (ABCF) program. The ABCF Program is funded by the Australian Department for Foreign Affairs and Trade (DFAT) through the BecA-CSIRO partnership; the Syngenta Foundation for Sustainable Agriculture (SFSA); the Bill & Melinda Gates Foundation (BMGF); the UK Department for International Development (DFID) and; the Swedish International Development Cooperation Agency (SIDA).

Conflicts of Interest: The authors declare no conflict of interest. The funders had no role in the design of the study; in the collection, analyses, or interpretation of data; in the writing of the manuscript, or in the decision to publish the results.

References

1. Minot, N.; Warner, J.; Lemma, S.; Kasa, L.; Gashaw, A.; Rashid, S. *The Wheat Supply Chain in Ethiopia: Patterns, Trends, and Policy Options*; International Food Policy Research Institute (IFPRI): Washington, DC, USA, 2015.
2. Sall, A.T.; Chiari, T.; Legesse, W.; Seid-Ahmed, K.; Ortiz, R.; van Ginkel, M.; Bassi, M.F. Durum wheat (*Triticum durum* Desf.): Origin, cultivation and potential expansion in sub-Saharan Africa. *Agronomy* **2019**, *9*, 263. [CrossRef]
3. Bona, L.; Wright, R.J.; Baligar, V.C. A rapid method for screening cereals for acid soil tolerance. *Cereal Res. Commun.* **1991**, *19*, 465–468.
4. Raman, H.; Ryan, P.R.; Raman, R.; Stodart, B.J.; Zhang, K.; Martin, P.; Wood, R.; Sasaki, T.; Yamamoto, Y.; Mackay, M.; et al. Analysis of *TaALMT1* traces the transmission of aluminum resistance in cultivated common wheat (*Triticum aestivum* L.). *Theor. Appl. Genet.* **2008**, *116*, 343–354. [CrossRef] [PubMed]
5. Aguilera, J.G.; Minozzo, J.A.D.; Barichello, D.; Fogaca, C.M.; da Silva, J.P.; Consoli, L.; Pereira, J.F. Alleles of organic acid transporter genes are highly correlated with wheat resistance to acidic soil in field conditions. *Theor. Appl. Genet.* **2016**, *129*, 1317–1331. [CrossRef] [PubMed]
6. Raman, H.; Stodart, B.; Ryan, P.R.; Delhaize, E.; Emebiri, L.; Raman, R.; Coombes, N.; Milgate, A. Genome-wide association analyses of common wheat (*Triticum aestivum* L.) germplasm identifies multiple loci for aluminium resistance. *Genome* **2010**, *53*, 957–966. [CrossRef] [PubMed]
7. Von Uexküll, H.R.; Mutert, E. Global extent, development and economic impact of acid soils. *Plant Soil* **1995**, *171*, 1–15. [CrossRef]
8. Wayima, E.F. Classification of Ethiopian soils with pH. *J. Soil Sci. Environ. Manag.* **2019**, in press.
9. Kidanemariam, A.; Gebrekidan, H.; Mamo, T.; Kibret, K. Impact of altitude and land use type on some physical and chemical properties of acidic soils in Tsegede highlands, northern Ethiopia. *Open J. Soil Sci.* **2012**, *2*, 223–233. [CrossRef]
10. Kidanemariam, A.; Gebrekidan, H.; Mamo, T.; Tesfaye, K. Wheat crop response to liming materials and N and P fertilizers in acidic soils of Tsegede highlands, northern Ethiopia. *Agric. For. Fish.* **2013**, *2*, 126–135. [CrossRef]
11. Mosissa, F. Progress of soil acidity management research in Ethiopia. *Greener J. Soil Sci. Plant Nutr.* **2018**, *5*, 9–22. [CrossRef]
12. Kochian, L.V.; Pineros, M.A.; Liu, J.P.; Magalhaes, J.V. Plant adaptation to acid soils: The molecular basis for crop aluminum resistance. *Ann. Rev. Plant Biol.* **2015**, *66*, 571–598. [CrossRef] [PubMed]
13. Magalhaes, J.V.; Garvin, D.F.; Wang, Y.H.; Sorrells, M.E.; Klein, P.E.; Schaffert, R.E.; Li, L.; Kochian, L.V. Comparative mapping of a major aluminum tolerance gene in sorghum and other species in the Poaceae. *Genetics* **2004**, *167*, 1905–1914. [CrossRef] [PubMed]
14. Bian, M.; Waters, I.; Broughton, S.; Zhang, X.Q.; Zhou, M.X.; Lance, R.; Sun, D.F.; Li, C.D. Development of gene-specific markers for acid soil/aluminium tolerance in barley (*Hordeum vulgare* L.). *Mol. Breed.* **2013**, *32*, 155–164. [CrossRef]
15. Ma, J.F.; Nagao, S.; Sato, K.; Ito, H.; Furukawa, J.; Takeda, K. Molecular mapping of a gene responsible for Al-activated secretion of citrate in barley. *J. Exp. Bot.* **2004**, *55*, 1335–1341. [CrossRef] [PubMed]
16. Ryan, P.R.; Raman, H.; Gupta, S.; Sasaki, T.; Yamamoto, Y.; Delhaize, E. The multiple origins of aluminium resistance in hexaploid wheat include *Aegilops tauschii* and more recent *cis* mutations to *TaALMT1*. *Plant J.* **2010**, *64*, 446–455. [CrossRef] [PubMed]
17. Foy, C.D. Tolerance of durum wheat lines to an acid, aluminum-toxic subsoil. *J. Plant Nutr.* **1996**, *19*, 1381–1394. [CrossRef]
18. Han, C.; Ryan, P.R.; Yan, Z.; Delhaize, E. Introgression of a 4D chromosomal fragment into durum wheat confers aluminium tolerance. *Ann. Bot.* **2014**, *114*, 135–144. [CrossRef]
19. Delhaize, E.; Ma, J.F.; Ryan, P.R. Transcriptional regulation of aluminium tolerance genes. *Trends Plant Sci.* **2012**, *17*, 341–348. [CrossRef]
20. Han, C.; Zhang, P.; Ryan, P.R.; Rathjen, T.M.; Yan, Z.H.; Delhaize, E. Introgression of genes from bread wheat enhances the aluminium tolerance of durum wheat. *Theor. Appl. Genet.* **2016**, *129*, 729–739. [CrossRef]
21. Garcia-Oliveira, A.L.; Martins-Lopes, P.; Tolra, R.; Poschenrieder, C.; Guedes-Pinto, H.; Benito, C. Differential physiological responses of Portuguese bread wheat (*Triticum aestivum* L.) genotypes under aluminium stress. *Diversity* **2016**, *8*, 26. [CrossRef]

22. Mengistu, D.K.; Kidane, Y.G.; Fadda, C.; Pe, M.E. Genetic diversity in Ethiopian durum wheat (*Triticum turgidum* var durum) inferred from phenotypic variations. *Plant Genet. Res. Charact. Util.* **2018**, *16*, 39–49. [CrossRef]

23. Mengistu, D.K.; Kiros, A.Y.; Pe, M.E. Phenotypic diversity in Ethiopian durum wheat (*Triticum turgidum* var. durum) landraces. *Crop J.* **2015**, *3*, 190–199. [CrossRef]

24. Mengistu, D.K.; Kidane, Y.G.; Catellani, M.; Frascaroli, E.; Fadda, C.; Pe, M.E.; Dell'Acqua, M. High-density molecular characterization and association mapping in Ethiopian durum wheat landraces reveals high diversity and potential for wheat breeding. *Plant Biotechnol. J.* **2016**, *14*, 1800–1812. [CrossRef] [PubMed]

25. Kabbaj, H.; Sall, A.T.; Al-Abdallat, A.; Geleta, M.; Amri, A.; Filali-Maltouf, A.; Belkadi, B.; Ortiz, R.; Bassi, F.M. Genetic diversity within a global panel of durum wheat (*Triticum durum*) landraces and modern germplasm reveals the history of alleles exchange. *Front. Plant Sci.* **2017**, *8*, 1277. [CrossRef] [PubMed]

26. Raman, H.; Hare, R.; Graham, K.; Coombes, N.; Raman, R. Characterisation of durum germplasm for aluminium resistance using nutrient solution culture. In *International Wheat Genetics Symposium*; Appels, R., Lagudah, R.E.E., Langridge, P., Mackay, M., McIntyre, L., Sharp, P., Eds.; Sydney University Press: Brisbane, Australia, 2008; pp. 1–3.

27. Joppa, L.R.; Williams, N.D. Langdon durum disomic substitution lines and aneuploid analysis in tetraploid wheat. *Genome* **1988**, *30*, 222–228. [CrossRef]

28. Delhaize, E.; Ryan, P.R.; Hebb, D.M.; Yamamoto, Y.; Sasaki, T.; Matsumoto, H. Engineering high-level aluminum tolerance in barley with the *ALMT1* gene. *Proc. Natl. Acad. Sci. USA* **2004**, *101*, 15249–15254. [CrossRef] [PubMed]

29. Zhou, G.F.; Delhaize, E.; Zhou, M.X.; Ryan, P.R. The barley *MATE* gene, *HvAACT1*, increases citrate efflux and Al^{+3} tolerance when expressed in wheat and barley. *Ann. Bot.* **2013**, *112*, 603–612. [CrossRef] [PubMed]

30. Bryan, G.J.; Dixon, A.; Gale, M.D.; Wiseman, G. A PCR-based method for the detection of hexaploid bread wheat adulteration of durum wheat and pasta. *J. Cereal Sci.* **1998**, *28*, 135–145. [CrossRef]

31. McNeil, D.; Lagudah, E.S.; Hohmann, U.; Appels, R. Amplification of DNA-sequences in wheat and its relatives—the Dgas44 and R350 families of repetitive sequences. *Genome* **1994**, *37*, 320–327. [CrossRef]

32. Zeven, A.C.; Waninge, J. The presence of 3 groups of Scalavatis and other hexaploid bread wheat plants contaminating durum-wheat fields in Cyprus. *Euphytica* **1989**, *43*, 117–124. [CrossRef]

33. Tovkach, A.; Ryan, P.R.; Richardson, A.E.; Lewis, D.C.; Rathjen, T.M.; Ramesh, S.; Tyerman, S.D.; Delhaize, E. Transposon-mediated alteration of *TaMATE1B* expression in wheat confers constitutive citrate efflux from root apices. *Plant Physiol.* **2013**, *161*, 880–892. [CrossRef]

34. Pereira, J.F.; Ryan, P.R. The role of transposable elements in the evolution of aluminium resistance in plants. *J. Exp. Bot.* **2019**, *70*, 41–54. [CrossRef] [PubMed]

35. Madlung, A.; Tyagi, A.P.; Watson, B.; Jiang, H.; Kagochi, T.; Doerge, R.W.; Martienssen, R.; Comai, L. Genomic changes in synthetic Arabidopsis polyploids. *Plant J.* **2005**, *41*, 221–230. [CrossRef] [PubMed]

36. Kidane, Y.G.; Mancini, C.; Mengistu, D.K.; Frascaroli, E.; Fadda, C.; Pe, M.E.; Dell'Acqua, M. Genome wide association study to identify the genetic base of smallholder farmer preferences of durum wheat traits. *Front. Plant Sci.* **2017**, *8*, 1230. [CrossRef] [PubMed]

37. Tsegaye, B.; Berg, T. Genetic erosion of Ethiopian tetraploid wheat landraces in Eastern Shewa, Central Ethiopia. *Genet. Resour. Crop Evol.* **2007**, *54*, 715–726. [CrossRef]

38. Pooniya, V.; Palta, J.A.; Chen, Y.; Delhaize, E.; Siddique, K.H. Impact of the *TaMATE1B* gene on above and below-ground growth of durum wheat grown on an acid and Al^{3+}-toxic soil. *Plant Soil* **2019**, 1–12. [CrossRef]

Article

Genetic Advance of Durum Wheat Under High Yielding Conditions: The Case of Chile

Alejandro del Pozo [1,*], Iván Matus [2], Kurt Ruf [1,2], Dalma Castillo [2], Ana María Méndez-Espinoza [1] and María Dolores Serret [3]

1   Centro de Mejoramiento Genético y Fenómica Vegetal, Facultad de Ciencias Agrarias, Universidad de Talca, 3460000 Talca, Chile
2   CRI-Quilamapu, Instituto de Investigaciones Agropecuarias, 3800062 Chillán, Chile
3   Centre de Recerca en Agrotecnologia (AGROTECNIO), 2519825198 Lleida, Spain
*   Correspondence: adelpozo@utalca.cl; Tel.: +56712200223

Received: 22 June 2019; Accepted: 13 August 2019; Published: 15 August 2019

Abstract: In Chile, durum wheat is cultivated in high-yielding Mediterranean environments, therefore breeding programs have selected cultivars with high yield potential in addition to grain quality. The genetic progress in grain yield (GY) between 1964 and 2010 was 72.8 kg ha^{-1} per year. GY showed a positive and significant correlation with days to heading, kernels per unit ground area and thousand kernel weight. The gluten and protein content tended to decrease with the year of cultivar release. The correlation between the δ^{13}C of kernels and GY was negative and significant (-0.62, $p < 0.05$, for all cultivars; and -0.97, $p < 0.001$, excluding the two oldest cultivars). The yield progress (genetic plus agronomic improvements) of a set of 40–46 advanced lines evaluated between 2006 and 2015 was 569 kg ha^{-1} per year. Unlike other Mediterranean agro-environments, a longer growing cycle together with taller plants seems to be related to the increase in the GY of Chilean durum wheat during recent decades.

Keywords: agronomic traits; carbon isotope; days to heading; grain quality; yield components

---

1. Introduction

Durum wheat (*Triticum turgidum* L. ssp. *durum*) covers ~17 million hectares worldwide, which is less than 10% of the total wheat area. However, its importance for human consumption is very high because it is used for making pasta, couscous, burghul and firik [1]. According to the International Grain Council, the largest producers of wheat in the world are the European Union, Canada, the United States, Turkey and Algeria.

For the production of high-quality durum wheat, dry environments are necessary, with warm days and cold nights during the growing season so that large grains are obtained with yellow color, vitreous kernels (more than 95%), hard texture and high test weight (about 82 kg hL^{-1}), alongside high protein content (greater than 10%) and strong gluten (greater than 30% wet gluten), which gives elasticity to dough for industrial use [2]. In Chile, durum wheat is grown in Mediterranean climate environments from the Valparaíso Region (32 °S) to the Biobio Region (37 °S), but mostly under irrigation conditions or in areas where rainfall is sufficient to satisfy most or all of the crop potential evapotranspiration. The sowing area has increased from 9600 ha in 2001 to 27,000 ha in 2015, and the average yield for 2011–2015 was 6.7 Mg ha^{-1} [3].

Wheat yields in different regions of the world have increased greatly since the 1960s as a result of genetic improvement and better agronomic practices [4]. With the Green Revolution, breeding programs have seen the introduction of semi-dwarfing genes that interfere with the action or production of gibberellin [5], leading to a reduction in plant size and an increase in the partitioning of the above-ground biomass towards spikes and grains [6,7]. In bread wheat, the genetic gain in grain yield (GY) was positively correlated with harvest index and the number of grains per spike and per m^2 [6,8]. In durum wheat, a Spanish study conducted with 12 cultivars from Italy and 12 from Spain released between 1930 and 2000 showed that the changes in grain yield were also associated with increases in the harvest index and number of grains per m^2 [9]. Another study carried out on 14 cultivars released in Italy between 1900 and 2000 indicated that the total aerial biomass had not changed and that the increase in yield was associated with a reduction in plant height and an increase in the harvest index and number of grains per m^2 [10]. These studies in durum wheat have been conducted under rain fed conditions and yields were below 6 Mg ha^{-1}, however, there is no information about the genetic progress in durum wheat in high yield potential environments (>10 Mg ha^{-1}) and how grain quality traits have been affected. Moreover, these works have not focused on studying exclusively the trends in breeding advances of post-Green-Revolution (i.e., semi-dwarf) durum wheat cultivars during the last half-century. This is despite the importance of this issue in the context of climate change and the fact that at least for bread wheat, there are studies reporting a stagnation in yields (or at least a drastic decrease in genetic advance) during the last decade [11].

In bread wheat, grain protein content, sedimentation value and wet gluten have increased in modern cultivars [8,12–14]. In durum wheat, modern dwarf and semi-dwarf cultivars have a higher gluten index compared to landraces or traditional Mediterranean cultivars [15]. Subira et al. [16] also reported significant changes in grain quality traits in a historical series of 24 durum wheat cultivars released in Italy and Spain in different periods of the 20th Century, particularly in gluten strength, sedimentation index and yellow color index. High protein content and 'strong' gluten are necessary to process semolina into a suitable final pasta product.

Physiological changes associated with breeding advances have also been reported for bread and durum wheat. For instance, modern bread wheat presented higher stomatal conductance (on an area basis) and carbon isotope discrimination (Δ^{13}C; or a lower carbon isotope composition, δ^{13}C), and lower oxygen isotope composition (δ^{18}O) than older varieties [8,17–19]. In durum wheat, modern varieties have higher Δ^{13}C (or lower δ^{13}C) compared to landraces [20–22], although no clear differences were found for δ^{18}O [21]. However, no information exists for durum wheat growing in a high-yielding Mediterranean environment.

The aim of this work was to analyze a) the changes in agronomic traits, grain quality and isotope composition in a set of ten durum wheat cultivars released in Chile between 1964 and 2010; and b) the progress in grain yield, plant height and test weight in selected advanced lines from the Instituto de Investigaciones Agropecuarias (INIA)-Chile breeding program. The experiments were conducted in a high-yielding Mediterranean environment between 2006 and 2015.

2. Materials and Methods

2.1. Experimental Site, Plant Material and Growing Conditions

The experiments were conducted at the Santa Rosa experimental field station (36°32′ S, 71°55′ W; 220 m.a.s.l.) of the Centro Regional de Investigación (CRI)-Quilamapu, Instituto de Investigaciones Agropecuarias (INIA). The climate corresponds to a humid Mediterranean type. During the experimental period (2006–2015), the monthly minimum average temperature was 3.1 °C (July) and the maximum 29.6 °C (January), and the average annual precipitation was 903 mm (Supplemental Table S1). The soil was a sandy loam, humic haploxerands (Andisol). Soil chemical characteristics of the top 10 cm were: pH 6.0, 8.87 mg kg^{-1} of N-N0$_3$; 17.05 mg kg^{-1} of P (Olsen), 0.45% of N-total, 4.5% of C and 0.33, 5.75, 0.65 and 0.48, cmol kg^{-1} of available K, Ca, Mg and Na, respectively [23].

Two different experiments were conducted. In the first experiment, ten cultivars released by the INIA breeding wheat program from 1964 to 2010 (Table 1) were evaluated during three consecutive seasons (2010 to 2012). The INIA cultivars derive from germplasm introduced from the International Maize and Wheat Improvement Center (CIMMYT) and probably all of them have the *Rht-B1* gene.

Table 1. Cultivars of durum wheat released by the wheat breeding program of the Chilean INIA between 1964 and 2010.

Cultivar	Year [1]	Cross/Pedigree
Alifén	1964	CAPELLI//ST 464
Quilafén	1970	YT54/Nl08//LD 357/2 *TC
Chagual INIA	1986	2156 3/AA" S"//PG" S"
Chonta INIA	1990	FRIGATTE"S"//RUFF/FLAMINGO"S"
Licán INIA	1990	RUFF "S"/FG"S"//MEX/3/SHWA"S"
Llareta INIA	1997	D67.54.4.9A//JORI'S'/ROSNER DURUM 119-200-4Y/3/ SAHEL77
Guayacán INIA	1997	ALTAR84/STINT"S"//SILVER
Corcolén INIA	2002	ALGA"S"/3/CANDEALFENS5/FLAMINGO"S"//PETREL"S"/ 4/CHURRILLA"S"/5/AUK"S"/6/RUFF"S"/FLAMINGO"S"// FLAMINGO"S"/CRANE"S"/3/YAVOROS 79/HUITLES"S"
Lleuque INIA	2009	YEL"S"/BAR"S"/3/GR"S"/AFN//CR"S"/5/DOM"S"//CR"S"*2/ GS"S"/3/SCO"S"/4/HORA/6/LAP76/GULL"S"/7/LICAN
QUC 3104–2005 [2]	2010	ALTAR84/ALD"S"//STN"S"/CHEN"S"/ALTAR84/4/ATES1D

[1] Year of cultivar release; [2] experimental line.

The experimental design was a complete block with four replications. Each plot consisted of five rows of 2.5 m length and 0.2 m apart. Sowing dates were in August of each year and the sowing rate was 220 kg ha^{-1}. Fertilization consisted of 1.5 t ha^{-1} of lime (88%–90% $CaCO_3$) before sowing, 260 kg ha^{-1} of diammonium phosphate (46% P_2O_5, 18% N), 200 kg ha^{-1} of potassium magnesium sulfate (22% K_2O, 18% MgO, 22% S), 90 kg ha^{-1} of potassium chloride (60% K_2O), 10 kg ha^{-1} of boronatrocalcite and 3 kg ha^{-1} of zinc sulfate (35% Zn) at sowing. After sowing, an extra 133 kg ha^{-1} of urea (46% of N) was applied at tillering initiation (Zadoks 20; [24]) and 201 kg ha^{-1} at the first node (Zadoks 31). Plots were furrow irrigated according to the needs of the crop (3–4 irrigations of ~50 mm each, per season). Weeds were controlled using the pre-emergence herbicide Bacara Forte 360SC, Bayer Crop Science (800 mL ha^{-1}; 12:12:12% w/v a.i. of flufenacet/flurtamone/diflufenican) and the post-emergence Ajax, Anasac, Chile (10 g ha^{-1}; 50% w/w a.i. of metsulfuron-methyl) and MCPA 750 SL, Anasac, Chile (800 mL ha^{-1}; 95% w/v a.i. of 2-methyl-4-chlorophenoxyacetic acid)). Since the oldest cultivars showed susceptibility to rust (*Puccinia striiformis* and *Puccinia triticina*), two applications were made of the foliar fungicide Juwel-Top, Basf (100 mL ha^{-1}; 12.5:12.5:15% a.i. of kresoxim-methyl/epoxiconazole/phenopropimorph). These applications were made before symptoms appeared, to avoid any interference of these diseases in the development of the plants.

In the second experiment, a selection of 46 advanced lines (F6–F8) of durum wheat from the breeding program (Durum Yield Nursery) and four check cultivars (Llareta-INIA, Corcolén-INIA, Lleuque-INIA and Queule-INIA) were tested each year from 2006 to 2015. Two trials of 25 genotypes each, including check cultivars, were established each year in an α-lattice design with five incomplete blocks per replicate, each block containing five genotypes. There were four replicates per genotype. The plots consisted of five rows of 2 m length and 0.20 m between rows. The seed rate was the equivalent of 220 kg ha^{-1}. The sowing date was August of each year. Crop fertilization and weed control were as recommended for each year. Plots were furrow irrigated according to crop need (3–4 irrigations of ~50 mm each, per season). These trials were regularly conducted by the breeding program in order to test the most promising advance lines in comparison with the commercial cultivars (check cultivars); those advance lines with outstanding performance were evaluated for more than one year, and the rest were replaced by new ones. As a consequence, the set of advance lines evaluated in each year was composed of different elite genotypes.

2.2. Agronomic Traits

In Experiment 1 the following traits were evaluated: (a) Days from emergence to heading (DH) through periodic observations (twice per week), when approximately half of the spikes in the plot had already extruded; (b) the number of spikes per m^2 (SM2) by counting the spikes in a 1.0 m length of a row; (c) the harvest index (HI), determined from a sample from the 1.0 m row at maturity and calculated as the ratio of grain dry weight to total above ground dry weight; (d) the number of kernels per spike (KS) and thousand kernel weight (TKW) from 25 spikes taken at random from each plot and (e) the number of kernels per m^2 (KM2) calculated as SM2 × KS. In Experiments 1 and 2, plant height (PH) from the ground to the top of the spike, excluding awns, was measured at maturity, and GY was assessed by harvesting 2 m^2 (five rows, 2 m long).

2.3. Grain Quality

The test weight was evaluated in Experiments 1 (2010) and 2 (2006 to 2015), in samples of wheat free of impurities (obtained from each genotype and replicate) using a 250 cc Schopper scale (Louis Schopper, Germany). In addition, grain samples obtained from the genotypes and replicates evaluated in Experiment 1 (in 2010) were ground in mill for wet gluten and protein content determination. Wet gluten content was determined according to the International Approved Methods of Analysis (AACCI Method 38–12.02) in 10 g of pure flour mixed with 5.5 mL of a 2% saline solution, which was homogenized and then placed in a gluten washer (Glutomatic® 2200, Perten Instruments, USA) for 5 min; then the wet gluten was weighed and expressed as a percentage of the amount of pure flour. Protein content (%) was also determined in ground grain samples placed in a quartz cuvette and the reflectance spectrum between 800 and 2500 nm was determined using near infrared reflectance spectroscopy (NIRS), Bruker, USA. Yellow berry incidence was assessed on 100 g of kernels, separating and weighing the affected grains and then expressed in percentage.

2.4. Total N Content and C and N Isotope Analyses

Measurements were performed in mature grains harvested in 2011 (Experiment 1). The total N content was analyzed using an elemental analyzer (Flash 1112 EA; ThermoFinnigan, Bremen, Germany). The stable carbon ($^{13}C/^{12}C$) and nitrogen ($^{15}N/^{14}N$) isotope ratios of the same mature grains were determined in the same elemental analyzer coupled with an isotope ratio mass spectrometer (Delta C IRMS, ThermoFinnigan, Bremen, Germany). Nitrogen was expressed as a concentration (g N per g of dry weight) and atropine was used as a system check in the elemental analyses of nitrogen. The $^{13}C/^{12}C$ ratios of plant material were expressed in δ notation: $\delta^{13}C = (^{13}C/^{12}C)$ sample/$(^{13}C/^{12}C)$ standard − 1, where 'sample' refers to plant material and 'standard' of known $^{13}C/^{12}C$ ratios. The $^{15}N/^{14}N$ ratios were also expressed in δ notation ($\delta^{15}N$) using international secondary standards of known $^{15}N/^{14}N$ ratios. More details are described in del Pozo et al. [8]. Measurements were performed at the Scientific Facilities of the University of Barcelona.

2.5. Data Analysis

Complete block analysis of variance (ANOVA) were performed for the set of cultivars evaluated in Experiment 1 using IBM SPSS Statistics software (SPSS Inc, USA). In addition, correlation analyses were performed between the year of cultivar release and agronomic, grain quality and isotope composition traits, and among the different traits. Trends for grain yield, plant height and test weight of 46 advanced lines and cultivars evaluated from 2005 to 2015 in Experiment 2 are also presented.

3. Results

3.1. Agronomic Traits in Cultivars Released During the Past Six Decades

Days to heading differed significantly among cultivars and also the year × cultivar interaction was significant (Table 2); it reduced in the 1990s, but then increased in the 2000s (Figure 1A). Plant height was significantly ($p < 0.001$) reduced from 108 cm in the 1960s to 90 cm in the 1970s, with a slight increase in 2010 (Table 2; Figure 1B).

Table 2. Mean sum of squares of the analysis of variance (ANOVA) for agronomic traits of ten durum wheat cultivars cultivated during three growing seasons (2010–2012).

Source of Variation	d.f.	DH	GY	PH	SM2	KS	KM2	TKW	HI
Year	2	238.9	187.9	418.1	413,645	420.6	345.5×10^6	63.5	0.017
Cultivar	9	84.2	20.3	456.7	36,213	287.0	50.4×10^6	292.5	0.019
Block	3	0.7	**9.0**	6.9	5219	16.3	19.5×10^6	0.2	0.001
Year × Cultivar	18	2.7	*3.0*	12.1	16,296	*29.6*	20.1×10^6	7.1	0.001
Residual	87	0.4	1.7	10.3	5040	11.5	12.8×10^6	2.9	0.001
Total	120								

Level of significance is indicated in bold ($p < 0.01$) and cursive ($p < 0.05$). DH: Days to heading; GY: Grain yield; PH: Plant height; SM2: Number of spikes per m^2; KS: Kernels per spike; KM2: Kernel number per m^2; TKW: Thousand kernel weight; HI: Harvest index.

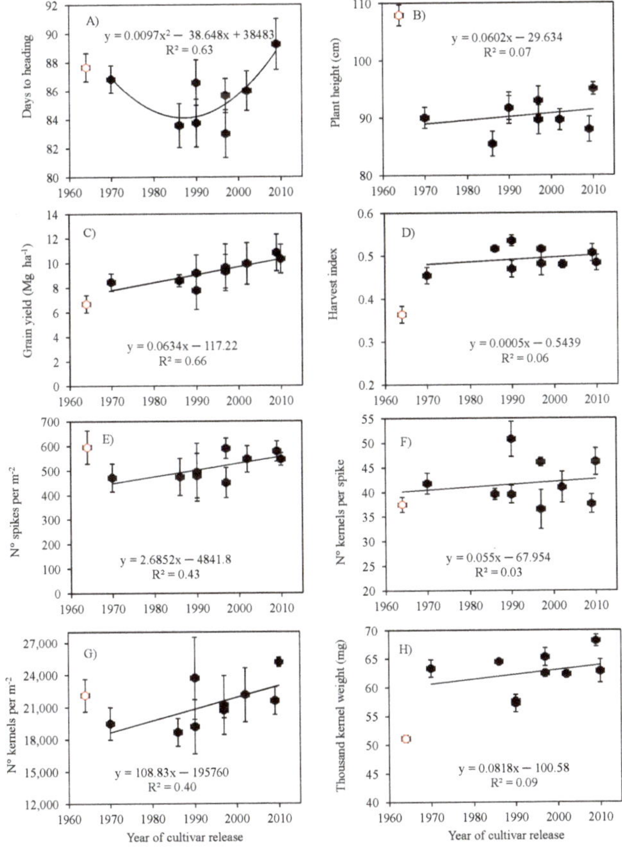

Figure 1. Relationships between the year of release of ten durum wheat cultivars and: Day to heading (**A**), plant height (**B**), grain yield (**C**), harvest index (**D**), number of spikes per square meter (**E**), number of kernels per spike (**F**), number of kernels per square meter (**G**) and thousand kernel weight (**H**). Values correspond to the average (±SE) of three growing seasons (2010–2012) except for HI, which was determined in 2010 and 2011. The oldest (1964) cultivar (open circle) was not considered in the regressions. Mean values of cultivars for each year of evaluation are shown in Supplemental Table S2.

GY exhibited a positive and linear relationship with the year of cultivar release ($R^2 = 0.66$; $p < 0.001$), this analysis excluding the oldest cultivar (1964; Figure 1C). The rate of increase in GY after 1960 was 72.8 kg ha^{-1} per year, and excluding the oldest cultivar it was 63.4 kg ha^{-1} per year. The SM2 of the ten cultivars ranged between 450 and 595 and increased significantly ($R^2 = 0.43$; $p < 0.05$) with the year of release (Figure 1E). The HI was 0.36 in the 1960s and increased to 0.45–0.53 in the 1970s and onwards, whereas TKW was 51.1 g in the 1960s and rose to 57–68 g after the 1970s, but neither trait was correlated with the year of cultivar release (Figure 1D,H). Similarly, KS was not correlated with the year of release (Figure 1F), but KM2 increased significantly ($R^2 = 0.40$; $p < 0.05$) with the year of cultivar release (Figure 1G).

The correlation matrix among the agronomic traits of the 10 cultivars evaluated during three growing seasons indicated that days to heading exhibited a positive and significant correlation with GY ($p < 0.05$) and KS ($p < 0.01$), and GY showed a positive and significant correlation with KM2 ($p < 0.05$) and TKW ($p < 0.001$; Table 3). Plant height was not correlated with GY. However, plant height showed a negative and highly significant ($p < 0.001$) correlation with TKW and HI. SM2 had a positive correlation with KM2 but a negative correlation with KS.

Table 3. Correlation matrix among agronomic traits evaluated in ten cultivars during three growing seasons (2010–2012).

	DH	GY	PH	SM2	KS	KM2	TKW	HI
DH	1.00							
GY	0.44 *	1.00						
PH	0.36	0.08	1.00					
SM2	−0.30	0.08	0.20	1.00				
KS	0.50 **	0.30	0.09	−0.61 ***	1.00			
KM2	0.06	0.39 *	0.33	0.76 ***	0.03	1.00		
TKW	0.14	0.59 ***	−0.55 ***	−0.03	0.01	0.01	1.00	
HI	−0.21	0.02	−0.75 ***	−0.32	0.34	−0.11	0.51 *	1.00

*: $p < 0.05$; **: $p < 0.01$; ***: $p < 0.001$ DH: Days to heading; GY: Grain yield; PH: Plant height; SM2: Spike number per m^2; KS: Kernels per spike; KM2: Kernel number per m^2; TKW: Thousand kernel weight; HI: Harvest index.

3.2. Grain Quality and Kernel Isotope Composition in Cultivars Released During the Past Six Decades

The test weight increased curvilinearly with the year of cultivar release (Figure 2A). The gluten and protein content tended to decrease with the year of cultivar release, although the correlations were not significant (Figure 2B,C). Yellow berry was higher in two cultivars, but there was no clear pattern with the year of cultivar release (Figure 2D).

The relationships between the year of cultivar release and N concentration or δ^{15}N in kernels were not significant (Figure 3A,B). The δ^{13}C of kernels tended to decrease with the year of cultivar release, although the correlation was not significant (Figure 3C). In addition, δ^{13}C was negatively correlated ($r = -0.62$; $p < 0.05$) with GY, but δ^{15}N was not correlated ($r = 0.03$; $p > 0.05$).

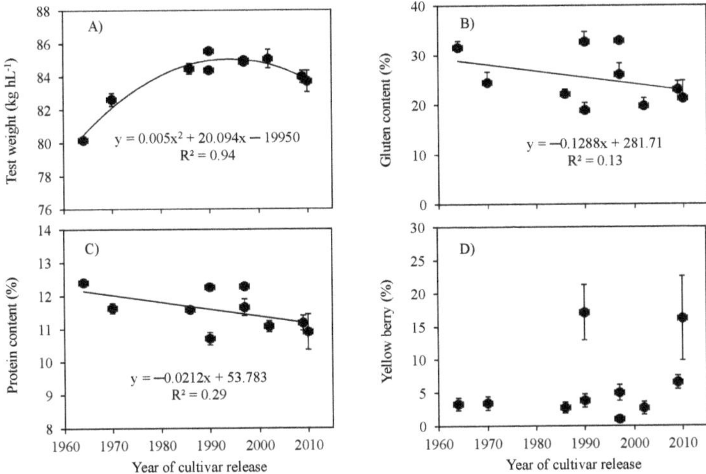

Figure 2. Relationships between the year of release of 10 durum wheat cultivars and kernel test weight (**A**), wet gluten content (**B**), protein content (**C**) and yellow berry (**D**), determined in 2010. Values correspond to the average (±SE) of four replicates.

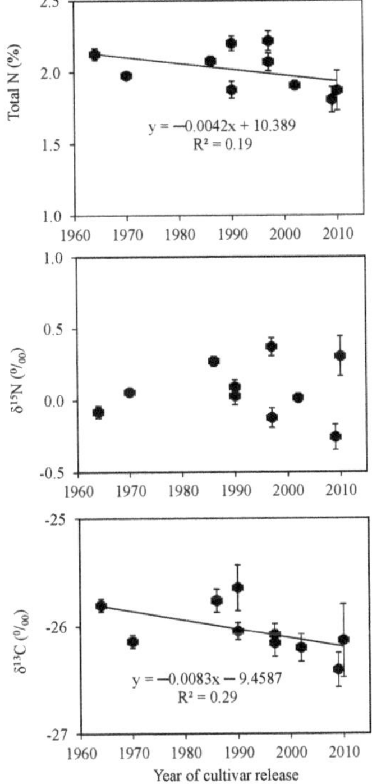

Figure 3. Relationships between the year of release of 10 durum wheat cultivars and the (**A**) total nitrogen, (**B**) natural abundance of ^{15}N (δ^{15}N) and (**C**) carbon isotope composition (δ^{13}C) in kernels, determined in 2011. Values correspond to the average (±SE) of four replicates.

3.3. Agronomic and Grain Quality Traits in Advanced Lines During the Last Decade

The GY and plant height of advanced lines increased from 2006 to 2015, reaching a maximum in 2011 with averages of 12.7 ± 0.8 Mg ha^{-1} and 96 ± 4.2 cm, respectively (Figure 4A,B). GY was highly correlated with plant height ($r = 0.85$; $p < 0.001$). The check cv. 'Corcolén' followed a similar trend to the advanced lines. The average GY of advanced lines and cultivars had a positive and significant ($R^2 = 0.50$; $p < 0.001$) relationship with the year of evaluation; the regression analysis indicated that the rate of increase in GY between 2006 and 2015 was 569 kg ha^{-1} per year. The test weight did not increase during this period (Figure 4C). No significant ($p < 0.05$) correlation was found between GY of advanced lines and the average temperature (maximum, minimum or mean) for the wheat growing season (August–January) from 2006 to 2015.

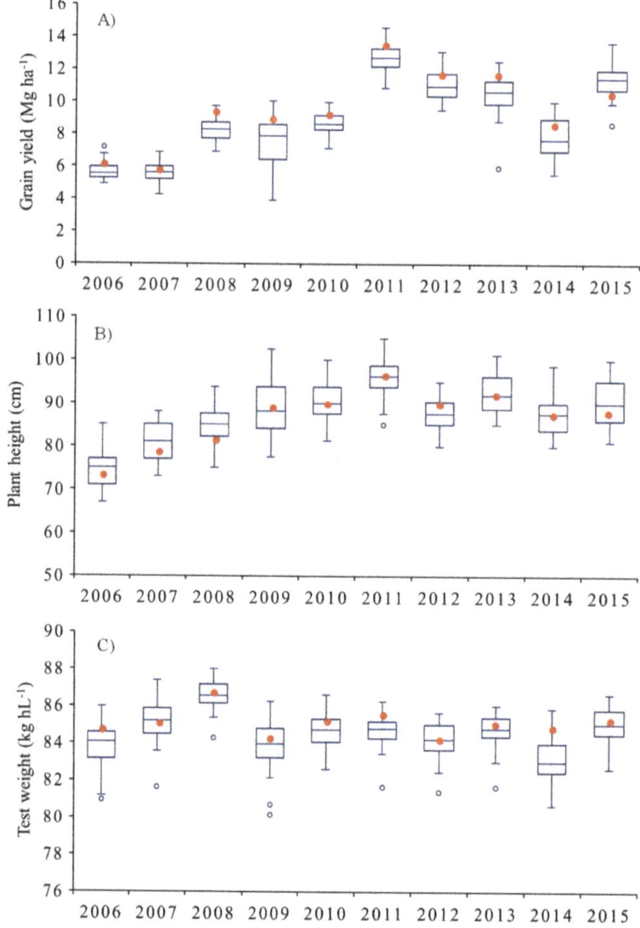

Figure 4. Grain yield (**A**), plant height (**B**) and test weight (**C**) for 40–46 advanced lines and cultivars of durum wheat grown under full irrigation in Santa Rosa, from 2006 to 2015. Box and whisker plots show the population minimum, 25th percentile/median/75th percentile and maximum. The open symbols indicate outlier data and the closed symbols indicate the check cultivar 'Corcolén'.

4. Discussion

4.1. Agronomic Traits

Modern cultivars of spring durum wheat from Chile have a very high yield potential (~13 Mg ha^{-1}) in a Mediterranean environment, under fully irrigated conditions. The yield potential achieved in Chile is clearly higher than values recorded in the Mediterranean basin. For example, high-yielding conditions in Spain usually do not surpass 8 Mg ha^{-1} [25,26], which is clearly lower than those achieved in the Mediterranean conditions of Chile. The high-yielding conditions in Spain usually imply several irrigations per season, particularly during the critical period from stem elongation to the middle grain filling, which alongside natural rainfall aims to balance the water lost due to accumulated evapotranspiration. Even so, the potential yields achieved in Spain are lower than in Chile due to a number of factors, such as Spain's shorter crop cycle duration, its higher night temperatures and the higher temperatures during the reproductive stage. The genetic advance in GY of spring durum wheat in the high-yielding environment of central Chile was 72.8 kg ha^{-1} per year (0.73% per year) for the period 1964–2010, and 63.4 kg ha^{-1} per year when the cultivar released in 1964 was excluded from the analysis (Figure 1). This is higher than the findings for spring bread wheat (43.5 kg ha^{-1} per year or 0.51% per year) for a similar period (1964–2008) and in the same Mediterranean environment [8]. It is also clearly higher than the increase reported for durum wheat in Spain (24 kg ha^{-1} y^{-1}; 0.44% y^{-1}) from 1980 to 2003, with no clear additional improvements occurring thereafter [26]. In northwest Mexico, under fully irrigated conditions, the genetic progress of spring durum and bread wheat varieties developed by CIMMYT was 0.49% and 0.41% per year, respectively, between 1966 and 2003 [27], and 0.88% per year when comparing eight bread wheat cultivars released between 1962 and 1988 [28]. A more recent study conducted at the same site in Mexico indicated that the GY progress was 30 kg ha^{-1} per year (0.59%) for spring bread wheat cultivars developed from 1966 to 2019 [29]. In Spain, under moderately irrigated conditions, the genetic progress of GY was 0.36% and 0.44% for Italian and Spanish cultivars of durum wheat, respectively, for cultivars released between 1920 and 2000 [9]. In South Australia, under rain fed conditions, the annual rate of increase in GY was 25 kg ha^{-1} for 13 cultivars released between 1958 and 2007 [30]. In North China, the annual genetic progress of spring bread wheat ranged from 0.48% (32.0 kg ha^{-1}) for cultivars released between the 1960s and the 1990s [31], and in Henan Province values of 51.3 kg ha^{-1} per year have been reported for the last three decades [32].

The yield progress observed in advanced lines of the INIA-Chile breeding program (Experiment 2), which includes genetic and agronomic progress, has been much higher (569 kg ha^{-1} per year) than in all the studies discussed above. This large increase in GY is explained partly by the genetic progress, but overall the improvements have derived from better agronomic management of durum wheat in the central-south of Chile, and this has included modifications to irrigation and particularly adjustments in fertilization practices conducted during the first three years of the program. In winter bread wheat, the yield progress was 246 kg ha^{-1} per year (2.6%) between 1976 and 1998 in central Chile under fully irrigated conditions [14]. Clearly, fine tuning of crop management can have large impacts on GY in high-yielding environments when lines or cultivars of high yield potential are available.

Plant height was reduced from 107 cm in 1964 to an average of 90 cm for the period 1970–2010 (Figure 1A), and this was the consequence of the introduction of semi-dwarfing genes in Chile in the late 1950s [33]. Plant height was negatively correlated with the year of release in Australia, in cultivars developed between 1958 and 1973, but not in cultivars released after 1973 [30], and in China, in cultivars released between 1960 and 2000 [34]. A negative correlation between plant height and GY was also reported in the study of Zhou et al. [34]. However, the comparison of advanced lines produced during the last decade (Figure 2) showed a positive correlation between plant height and GY. These results suggest that plant height of semi-dwarf wheat below 70–80 cm may limit light interception and thus canopy photosynthesis and yield potential in high-yielding environments.

The HI increased between 1964 and 1970, but after that there were no changes. Furthermore, the correlation between GY and HI was not significant. The maximum values of harvest index (0.53) found in the current work were higher than those reported by Royo et al. [9] in a set of Italian and Spanish cultivars of durum wheat released between 1920 and 2000 and tested in Spain. In studies where cultivars released before and after the green revolution were evaluated, HI and the year of cultivar release were positively correlated (e.g., [9] for durum; [6,8,31] for bread wheat), but there was no correlation in cultivars released after 1970 (Figure 1C; see also [29]).

The increase in GY was positively associated with days to heading and KM2 and TKW (Table 3). The increase in the crop cycle in an irrigated Mediterranean environment contrasts with the breeding trend observed in rain fed Mediterranean areas, where early flowering, shorter duration cultivars are selected to escape post anthesis drought [9,35,36].

TKW increased significantly from 1964 to 1970, but the correlation with the year of cultivar release was not significant for the period 1970–2000 (Figure 1F). Genetic progress in TKW can be positive, negative or null depending on whether kernel weight has been a selection target for breeders and whether there have been changes in the number of grains per year (the trade-off between seed size and number in crops; [37]). For instance, in durum wheat growing in Mediterranean environments, kernel weight was superior in modern cultivars in Turkey [38], but remained unchanged in Italian and Spanish cultivars from the 20th century [9]. In bread wheat, kernel weight has been reduced [8,14,39] or has not changed [31] with genetic improvement.

4.2. Grain Quality and Kernel Isotope Composition

The test weight increased in modern cultivars and was positively correlated with TKW ($r = 0.44$; $p < 0.05$). The values of test weight obtained in this work are higher than those found in durum wheat genotypes grown under rain fed conditions in different zones of Spain [40,41]. Unfortunately, gluten and protein content did not improve between 1964 and 2010. Other studies comparing older cultivars or landraces to modern cultivars of durum wheat from Mediterranean countries have revealed lower grain nitrogen or protein content in the more modern cultivars [15,16,20,21,42]. In addition, the presence of *Rht* dwarfing genes in bread and durum wheat seems to reduce the concentration of Zn, Fe, Mn and Mg in kernels [43].

A number of studies have reported a negative correlation between grain protein concentration and GY in durum wheat [44] and in bread wheat [45,46]. It is probable that the lack of genetic progress in protein content is related to the strong increase in GY of the Chilean cultivars. However, this negative relationship should not be a limitation for genetic improvement in quality traits in grains of durum wheat because protein composition seems to be more important than the concentration [16,47].

The relationship between kernel δ^{13}C and GY was negative, suggesting that genotypes exhibiting higher water use are the most productive [21,48]. In bread wheat under fully irrigated conditions, modern and more productive cultivars showed lower δ^{18}O and δ^{13}C, and higher stomatal conductance [8,18,21,25]. This negative relationship between δ^{13}C and GY (or positive relationship between Δ^{13}C and GY) has also been found in rain fed Mediterranean conditions ([21] for durum wheat; [49,50] for bread wheat), suggesting that the most productive lines are those able to maintain higher stomatal conductance and use more water [51]. In addition, the stomatal conductance of post green revolution wheat cultivars in Australia seem to show a lower sensitivity to vapor pressure deficit above 2 kPa compared to older cultivars [52], and this can be associated with lower (more negative) δ^{13}C values.

In summary, changes in a number of traits have occurred in durum wheat cultivars selected for high-yielding environments in Chile. The large genetic progress in grain yield was associated with increases in days to heading, KM2 and TKW. The test weight has also increased with the year of cultivar release, but the gluten and protein content have not improved between 1964 and 2010. Interestingly, the increase in yield potential seems related to longer duration and somewhat taller plants that are able to use more water.

5. Conclusions

This study provided evidence that a high genetic advance in GY for durum wheat is feasible under high yielding conditions. The increase in GY was a consequence of a greater number of kernels per m^2 and higher kernel weight in the more modern cultivars. The test weight was lower in the 1960s and increased curvilinearly with year of cultivar release. The gluten and protein content did not improve between 1964 and 2010. GY was negatively correlated with kernel $\delta^{13}C$, suggesting that genotypes exhibiting higher water use are the most productive. The yield progress of a set of advanced lines evaluated between 2006 and 2015 was very high, due to genetic progress, but this was also due to management improvements, particularly adjustment of fertilization practices conducted during the first three years. Unlike other Mediterranean agro-environments, a longer growing cycle together with taller plants seems to be related to the increase in the GY of Chilean durum wheat during recent decades.

Supplementary Materials: The following are available online at http://www.mdpi.com/2073-4395/9/8/454/s1, Table S1: Monthly minimum (T min) and maximum (T max) temperatures and precipitation (PP) at Santa Rosa, Table S2: Mean values of cultivar traits according to the year of evaluation.

Author Contributions: I.M. designed the experiment and selected the germplasm. D.C. and K.R. were in charge of the management of the experiment and evaluation of agronomic traits. M.D.S. contributed to the isotope analysis. A.M.M.-E. contributed to data analysis. A.d.P. performed the data analysis and was in charge of writing the text but all the authors contributed to the manuscript.

Funding: This work was supported by the research CONICYT grant FONDECYT N° 1180252, Chile, and the contribution of Maria Dolores Serret was supported in part by the AGL2016-76527-R project from MINECO, Spain.

Acknowledgments: We thank Alejandro Castro for technical assistance with the field experiments.

Conflicts of Interest: The autors declare no conflict of interest.

References

1. Elias, E.M.; Manthey, F.A. End products: Present and future uses. In *Durum Wheat Breeding Current Approaches and Future Strategies*; Royo, C., Nachit, M.M., di Fonzo, N., Araus, J.L., Pfeiffer, W.H., Slafer, G.A., Eds.; Food Products Press: New York, NY, USA, 2005; pp. 63–86.

2. Acevedo, E.; Silva, P. Trigo candeal: Calidad, mercado y zonas de cultivo. In *Serie Ciencias Agronómicas N° 12*; Universidad de Chile, Facultad de Ciencias Agronómicas: Santiago, Chile, 2007. (In Spanish)

3. ODEPA. Estadísticas Productivas. 2017. Available online: http://www.odepa.gob.cl/estadisticas/productivas/ (accessed on 23 November 2018). (In Spanish).

4. Miralles, D.J.; Slafer, G.A. Sink limitations to yield in wheat: How could it be reduced? *J. Agric. Sci.* **2007**, *145*, 139–149. [CrossRef]

5. Hedden, P. The genes of the Green Revolution. *Trends Genet.* **2003**, *19*, 5–9. [CrossRef]

6. Shearman, V.J.; Sylvester-Bradley, R.; Scott, R.K.; Foulkes, M.J. Physiological processes associated with wheat yield progress in the UK. *Crop Sci.* **2005**, *45*, 175–185.

7. Del Pozo, A.; Matus, I.; Araus, J.L.; Serret, D. Agronomic and physiological traits associated with breeding advances of wheat under high-productive Mediterranean conditions. The case of Chile. *Environ. Exp. Bot.* **2014**, *103*, 180–189. [CrossRef]

8. Trethowan, R.M.; Reynolds, M.P.; Ortiz-Monasterio, J.I.; Ortiz, R. The genetic basis of the green revolution in wheat production. *Plant Breed. Rev.* **2007**, *8*, 39–58.

9. Royo, C.; Álvaro, F.; Martos, V.; Ramdani, A.; Isidro, J.; Villegas, D.; García del Moral, L.F. Genetic changes in durum wheat yield components and associated traits in Italian and Spanish varieties during the 20th century. *Euphytica* **2007**, *155*, 259–270. [CrossRef]

10. De Vita, P.; Li Destri Nicosia, O.; Nigro, F.; Platani, C.; Riefolo, C.; di Fonzo, N.; Cattivelli, L. Breeding progress in morpho-physiological, agronomical and qualitative traits of durum wheat cultivars released in Italy during the 20th century. *Eur. J. Agron.* **2007**, *26*, 39–53. [CrossRef]

11. Lo Valvo, P.J.; Miralles, D.J.; Serrago, R.A. Genetic progress in Argentine bread wheat varieties released between 1918 and 2011: Changes in physiological and numerical yield components. *Field Crop Res.* **2018**, *221*, 314–321. [CrossRef]

12. Ortiz-Monasterio, J.I.; Peña, R.J.; Sayre, K.D.; Rajaram, S. CIMMYT's Genetic Progress in Wheat grain quality under four nitrogen rates. *Crop Sci.* **1997**, *37*, 892–898. [CrossRef]
13. Dencic, S.; Kobiljski, B.; Mladenov, N.; Hristov, N.; Pavlovic, M. Long-term breeding for bread making quality in wheat. In *Wheat Production in Stressed Environments*; Buck, H.T., Nisi, J.E., Salomon, N., Eds.; Springer: New York, NY, USA, 2005; pp. 495–501.
14. Matus, I.; Mellado, M.; Pinares, M.; Madariaga, R.; del Pozo, A. Genetic progress in winter wheat cultivars released in Chile from 1920 and 2000. *Child. J. Agric. Res.* **2012**, *72*, 303–308. [CrossRef]
15. Motzo, R.; Fois, S.; Giunta, F. Relationship between grain yield and quality of durum wheats from different eras of breeding. *Euphytica* **2004**, *140*, 147–154. [CrossRef]
16. Subira, J.; Peña, R.J.; Álvaro, F.; Ammar, K.; Ramdani, A.; Royo, C. Breeding progress in the pasta-making quality of durum wheat cultivars released in Italy and Spain during the 20th Century. *Crop Pasture Sci.* **2014**, *65*, 16–26. [CrossRef]
17. Fischer, R.A.D.; Rees, K.D.; Sayre, Z.M.; Lu, A.G.; Condon, A. Larqué-Saavedra, Wheat yield progress associated with higher stomatal conductance and photosynthetic rate, and cooler canopies. *Crop Sci.* **1998**, *38*, 1467–1475. [CrossRef]
18. Barbour, M.M.; Fischer, R.A.; Sayre, K.D.; Farquhar, G.D. Oxygen isotope ratio of leaf and grain material correlates with stomatal conductance and grain yield in irrigated wheat. *Aust. J. Plant Physiol.* **2000**, *27*, 625–637. [CrossRef]
19. Foulkes, M.J.; de Silva, J.; Gaju, O.; Carvalho, P. Relationships between δ^{13}C, δ^{18}O and grain yield in bread wheat genotypes under favourable irrigated and rain-fed conditions. *Field Crop Res.* **2016**, *196*, 237–250. [CrossRef]
20. Araus, J.L.; Ferrio, J.P.; Buxó, R.; Voltas, J. The historical perspective of dryland agriculture: Lessons learned from 10,000 years of wheat cultivation. *J. Exp. Bot.* **2007**, *58*, 131–145. [CrossRef] [PubMed]
21. Araus, J.L.; Cabrera-Bosquet, L.; Serret, M.D.; Bort, J.; Nieto-Taladriz, M.T. Comparative performance of δ^{13}C, δ^{18}O and δ^{15}N for phenotyping durum wheat adaptation to a dryland environment. *Funct. Plant Biol.* **2013**, *40*, 595–608. [CrossRef]
22. Royo, C.; Martos, V.; Ramdani, A.; Villegas, D.; Rharrabti, Y.; García del Moral, L.F. Changes in yield and carbon isotope discrimination of Italian and Spanish durum wheat during the 20th century. *Agron. J.* **2008**, *100*, 352–360. [CrossRef]
23. Zagal, E.; Rodríguez, N.; Vidal, I.; Quezada, L. Actividad microbiana en un suelo de origen volcánico bajo distinto manejo agronómico. *Agric. Técnica* **2002**, *62*, 297–309. (In Spanish) [CrossRef]
24. Zadoks, J.C.; Chang, T.T.; Konzak, C.F. A decimal code for the growth stages of cereals. *Weed Res.* **1974**, *14*, 415–421. [CrossRef]
25. Araus, J.L.; Villegas, D.; Aparicio, N.; García del Moral, L.F.; El Hani, S.; Rharrabti, Y.; Ferrio, J.P.; Royo, C. Environmental factors determining carbon isotope discrimination and yield in durum wheat under Mediterranean conditions. *Crop Sci.* **2003**, *43*, 170–180. [CrossRef]
26. Chairi, F.; Vergara Diaz, O.; Vatter, T.; Aparicio, N.; Nieto-Taladriz, M.T.; Kefauver, S.C.; Bort, J.; Serret, M.D.; Araus, J.L. Post-Green Revolution genetic advance in durum wheat: The case of Spain. *Field Crop Res.* **2018**, *228*, 158–169. [CrossRef]
27. Fischer, R.A. Understanding the physiological basis of yield potential in wheat. *J. Agric. Sci.* **2007**, *145*, 99–113. [CrossRef]
28. Sayre, K.D.; Rajaram, S.; Fischer, R.A. Yield progress in short bread wheats in northwest Mexico. *Crop Sci.* **1997**, *37*, 36–42. [CrossRef]
29. Aisawi, K.A.B.; Reynolds, M.P.; Singh, R.P.; Foulkes, M.J. The physiological basis of the genetic progress in yield potential of CIMMYT spring wheat cultivars from 1966 to 2009. *Crop Sci.* **2015**, *55*, 1749–1764. [CrossRef]
30. Sadras, V.; Lawson, C. Genetic gain in yield and associated changes in phenotype, trait plasticity and competitive ability of South Australian wheat varieties released between 1958 and 2007. *Crop Pasture Sci.* **2011**, *62*, 533–549. [CrossRef]
31. Zhou, Y.; He, Z.H.; Chen, X.M.; Wang, D.S.; Yan, J.; Xia, X.C.; Zhang, Y. Genetic improvement of wheat yield potential in North China. In *Wheat Production in Stressed Environments*; Buck, H.T., Nisi, J.E., Salomón, N., Eds.; Springer: New York, NY, USA, 2007; pp. 583–589.

32. Zheng, T.C.; Zhang, X.K.; Yin, G.H.; Wang, L.N.; Han, Y.L.; Chen, L.; Huang, F.; Tang, J.W.; Xia, X.C.; He, Z.H. Genetic gains in grain yield, net photosynthesis and stomatal conductance achieved in Henan Province of China between 1981 and 2008. *Field Crop Res.* **2011**, *122*, 225–233. [CrossRef]

33. Zapata, C.; Silva, P.; Acevedo, E. Comportamiento de isolíneas de altura en relación con el rendimiento y distribución de asimilados en trigo. *Agric. Técnica* **2004**, *64*, 139–155. (In Spanish) [CrossRef]

34. Zhou, B.; Elazab, A.; Bort, J.; Sanz-Sáez, A.; Nieto-Taladriz, M.T.; Serret, M.D.; Araus, J.L. Agronomic and physiological responses of Chinese facultative wheat genotypes to high-yielding Mediterranean conditions. *J. Agric. Sci.* **2016**, *154*, 870–889. [CrossRef]

35. Perry MWD'Antuono, M.F. Yield improvement and associated characteristics of some Australian spring wheat cultivars introduced between 1860 and 1982. *Aust. J. Agric. Res.* **1989**, *40*, 457–472.

36. Loss, S.P.; Kirby, E.J.M.; Siddique, K.H.M.; Perry, M.W. Grain growth and development of old and modern Australian wheats. *Field Crop Res.* **1989**, *21*, 131–146. [CrossRef]

37. Sadras, V.O. Evolutionary aspects of the trade-off between seed size and number in crops. *Field Crop Res.* **2007**, *100*, 125–138. [CrossRef]

38. Koç, M.; Celaleddin, B.; Genç, I. Photosynthesis and productivity of old and modern durum wheats in a Mediterranean environment. *Crop Sci.* **2003**, *43*, 2089–2098. [CrossRef]

39. Guarda, G.; Padovan, S.; Delogu, G. Grain yield, nitrogen-use efficiency and baking quality of old and modern Italian bread-wheat cultivars grown at different nitrogen levels. *Eur. J. Agron.* **2004**, *21*, 181–192. [CrossRef]

40. Rharrabti, Y.; Royo, C.; Villegas, D.; Paricio, N.; García del Moral, L.F. Durum wheat quality in Mediterranean environments: I. Quality expression under different zones, latitudes and water regimes across Spain. *Field Crop Res.* **2003**, *80*, 123–131. [CrossRef]

41. Garrido-Lestache, E.; López-Bellido, R.J.; López-Bellido, L. Durum wheat quality under Mediterranean conditions as affected by N rate, timing and splitting, N form and S fertilization. *Eur. J. Agron.* **2005**, *23*, 265–278. [CrossRef]

42. Acreche, M.M.; Slafer, G.A. Variation of grain nitrogen content in relation with grain yield in old and modern Spanish wheats grown under a wide range of agronomic conditions in a Mediterranean region. *J. Agric. Sci.* **2009**, *147*, 657–667. [CrossRef]

43. Velu, G.; Singh, R.P.; Huerta, J.; Guzmán, C. Genetic impact of *Rht* dwarfing genes on grain micronutrients concentration in wheat. *Field Crop Res.* **2017**, *214*, 373–377. [CrossRef]

44. Blanco, A.; Mangini, G.; Giancaspro, A.; Giove, S.; Colasuonno, P.; Simeone, R.; Signorile, A.; De Vita, P.; Mastrangelo, A.M.; Cattivelli, L. Relationships between grain protein content and grain yield components through quantitative trait locus analyses in a recombinant inbred line population derived from two elite durum wheat cultivars. *Mol. Breed.* **2012**, *30*, 79–92. [CrossRef]

45. Oury, F.X.; Godin, C. Yield and grain protein concentration in bread wheat: How to use the negative relationship between the two characters to identify favourable genotypes? *Euphytica* **2007**, *157*, 45–57. [CrossRef]

46. Bogard, M.; Allard, V.; Brancourt-Hulmel, M.; Heumez, E.; Machet, J.M.; Jeuffroy, M.-H.; Gate, P.; Martre, P.; Le Gouis, J. Deviation from the grain protein concentration–grain yield negative relationship is highly correlated to post-anthesis N uptake in winter wheat. *J. Exp. Bot.* **2010**, *61*, 4303–4312. [CrossRef] [PubMed]

47. De Santis, M.A.; Giuliani, M.M.; Giuzio, L.; de Vita, P.; Lovegrove, A.; Shewry, P.R.; Flagella, Z. Differences in gluten protein composition between old and modern durum wheat genotypes in relation to 20th century breeding in Italy. *Eur. J. Agron.* **2017**, *87*, 19–29. [CrossRef] [PubMed]

48. Blum, A. Effective use of water (EUW) and not water-use efficiency (WUE) is the target of crop yield improvement under drought stress. *Field Crop Res.* **2009**, *112*, 119–123. [CrossRef]

49. Rebetzke, G.J.; Condon, A.G.; Farquhar, G.D.; Appels, R.; Richards, R.A. Quantitative trait loci for carbon isotope discrimination are repeatable across environments and wheat mapping populations. *Appl. Genet.* **2008**, *118*, 123–137. [CrossRef] [PubMed]

50. Del Pozo, A.; Yáñez, A.; Matus, I.; Tapia, G.; Castillo, D.; Araus, J.L.; Sanchez-Jardón, L. Physiological traits associated with wheat yield potential and performance under water-stress in a Mediterranean environment. *Front. Plant Sci.* **2016**, *7*, 987. [CrossRef] [PubMed]

51. Roche, D. Stomatal conductance is essential for higher yield potential of C3 crops. *Crit. Rev. Plant Sci.* **2015**, *34*, 429–453. [CrossRef]

52. Schoppach, R.; Fleury, D.; Sinclair, T.R.; Sadok, W. Transpiration sensitivity to evaporative demand across 120 years of breeding of Australian wheat cultivars. *J. Agric. Crop Sci.* **2017**, *203*, 219–226. [CrossRef]

Article

Durum Wheat Seminal Root Traits within Modern and Landrace Germplasm in Algeria

Ridha Boudiar [1,2,*], Juan M. González [3], Abdelhamid Mekhlouf [1], Ana M. Casas [4] and Ernesto Igartua [4,*]

1   Laboratoire d'Amélioration et de Développement de la Production Végétale et Animale (LADPVA), University of Ferhat ABBAS (UFAS-Sétif1), Setif 19000, Algeria; mekhloufa@yahoo.fr
2   Biotechnology Research Center (CRBt), UV 03 BP E73, Nouvelle Ville Ali Mendjli, Constantine 25016, Algeria
3   Department of Biomedicine and Biotechnology, University of Alcalá, 28805 Alcalá de Henares, Spain; juanm.gonzalez@uah.es
4   Estación Experimental de Aula Dei, EEAD-CSIC, Avenida Montañana 1005, 50059 Zaragoza, Spain; acasas@eead.csic.es
*   Correspondence: boudiarreda@yahoo.fr (R.B.); igartua@eead.csic.es (E.I.); Tel.: +213-(31)-775037/39 (R.B.); +34-976716092 (E.I.)

Received: 11 March 2020; Accepted: 14 May 2020; Published: 16 May 2020

Abstract: Seminal roots are known to play an important role in crop performance, particularly under drought conditions. A set of 37 durum wheat cultivars and local landraces was screened for variation in architecture and size of seminal roots using a laboratory setting, with a filter paper method combined with image processing by SmartRoot software. Significant genetic variability was detected for all root and shoot traits assessed. Four rooting patterns were identified, with landraces showing overall steeper angle and higher root length, in comparison with cultivars, which presented a wider root angle and shorter root length. Some traits revealed trends dependent on the genotypes' year of release, like increased seminal root angle and reduced root size (length, surface, and volume) over time. We confirm the presence of a remarkable diversity of root traits in durum wheat whose relationship with adult root features and agronomic performance should be explored.

Keywords: proxy traits; genetic resources; root screening; root architecture

---

1. Introduction

The root system of wheat includes two main types, seminal (embryonic) and nodal roots, also known as the crown or adventitious roots [1,2]. Both types of roots play a crucial role in plant growth and are active throughout the whole plant life. Seminal roots, however, could be more important under specific circumstances, like drought conditions, as they penetrate deeper into the soil layers than nodal roots, making water in deep layers accessible to the plant [3–5]. Seminal roots also play a capital role during crop establishment, as they are the only roots existing before the emergence of the fourth leaf. Seminal roots include one primary root, two pairs of symmetric roots at each side, and, at times, a sixth central root [6].

The main features of root systems are encompassed under two categories, root system architecture (RSA) and morphology. RSA is related to the whole, or a large subset, of the root system, and may be described as topological or geometric measures of the root shape. Root morphology, as defined by J. Lynch, refers to "the surface features of a single root axis as an organ, including characteristics of the epidermis such as root hairs, root diameter, the root cap, the pattern of appearance of daughter roots, undulations of the root axis, and cortical senescence" [7]. The traits often used to describe wheat roots are total root length, root surface area, root volume, root angle, number of roots, and root diameter [8–10].

Roots are difficult to measure readily in natural conditions. Root trait determination has become accessible through the development of phenotyping methods in artificial systems, for instance, gel chambers [11], rolled germination paper [12], clear pots and growth pouches [13], "Termita" chamber and Whatman paper system [14], or growth pouches system [15]. Seminal roots can be phenotyped early and easily compared to the root system of mature plants [16,17], and for this reason, they have been proposed as good candidates to act as proxy traits in wheat [18] and maize [19,20]. Nevertheless, phenotyping these traits could be of interest only if they are useful to predict root growth and functioning in adult plants [21,22]. Indeed, several studies have found useful associations with traits in adult plants of wheat species [23–26]. For instance, the seminal root angle was correlated with nodal root angle [5,27], and with grain yield under drought conditions [28]. The seminal root number was correlated with thousand kernel weight (TKW) under stress, while the primary root length at the seedling stage was correlated with TKW under wetter conditions [25]. A steeper angle between the outermost roots and a higher root number in wheat seedlings have been linked to a more compact root system with more roots at depth in wheat [11,24,29].

Genotypic variation in root architecture has been reported within genotypes of different crop species [30–32], including wheat [13,21,25,33]. The presence of variation for the trait of interest is an essential requirement to improve the adaptability of crops under changing environmental conditions [34].

Local landraces are considered well adapted to the region where they were grown and contain large genetic diversity useful to improve crops like durum wheat [35]. These landraces were replaced by high yielding but more uniform semi-dwarf cultivars, better adapted to modern agriculture. However, scientists are convinced that local landraces still constitute a genetic resource useful to improve commercially valuable traits [36]. It is assumed that root traits enhance response to drought stress [37], but the realization of their contribution to superior grain yield depends on the type of drought and the agro-ecological conditions [38]. A deep rooting ideotype ("steep, cheap, and deep") was proposed by [39] to optimize water and N acquisition, building on the assumption that deeper rooting genotypes will use water that is beyond reach for shallower rooting genotypes. Modern breeding has caused some shifts in the root system architecture of durum wheat, from shallower and densely rooted systems in landraces of Mediterranean origin to deeper and more evenly distributed systems throughout the soil depth in cultivars worldwide [5].

The current study aims at evaluating the diversity of seminal root traits, including root angle and depth, during early growth of a set of durum wheat genotypes, consisting of modern cultivars and local landraces which are representative of the germplasm adapted to the mostly semi-arid conditions of Algerian cereal-growing regions before and after the advent of modern breeding. The study aims to reveal morphological diversity that could have agronomic relevance and, therefore, interest breeders.

2. Materials and Methods

2.1. Plant Material

We studied thirty-seven genotypes (landraces and modern cultivars), representative of durum wheat (*Triticum turgidum* ssp. *durum* Desf.) grown in Algeria. Geographical origins were varied (Algeria, France, Italy, Spain, Tunisia), and included genotypes produced at international breeding programs addressing semi-arid areas, namely the International Maize and Wheat Improvement Center (CYMMIT), the International Center for Agricultural Research in the Dry Area (ICARDA) and the Arab Center for the Studies of Arid zones and Drylands (ACSAD). These genotypes are representative of different periods of agriculture in Algeria, before and after the Green Revolution (Table 1).

Table 1. Name, type of cultivar, origin, and year of release of 37 genotypes of durum wheat used in the experiment.

N°	Genotype	Type/Pedigree	Origin	Year of Release
1	Beliouni	Landrace	Algeria	1958
2	Bidi 17	Landrace	Algeria	1930
3	Djenah Khotifa	Landrace	North Africa	1955
4	Gloire de Montgolfier	Landrace	Algeria	1960
5	Guemgoum R' khem	Landrace	Algeria	1960
6	Hedba 3	Landrace	Algeria	1921
7	Langlois	Landrace	Algeria	1930
8	Mohammed Ben Bachir (MBB)	Landrace	Algeria	1930
9	Montpellier	Landrace	Algeria	1965
10	Oued Zenati 368	Landrace	Algeria	1936
11	Acsad 65	Gerardo-vz-469/3/Jori-1//Nd-61-130/Leeds	ACSAD	1984
12	Altar 84	Ruff/Flamingo,mex//Mexicali-75/3/Shearwater	CYMMIT	1984
13	Ammar 6	Lgt3/4/Bicre/3/Ch1// Gaviota/Starke	ICARDA	2010
14	Bousselem	Heider//Martes/ Huevos de oro	ICARDA	2007
15	Boutaleb	Hedba 3/Ofanto	Algeria	2013
16	Capeiti	Eiti*6/Senatore-Cappelli	Italy	1940
17	Chen's	Shearwater(sib)/(sib)Yavaros-79	CYMMIT	1983
18	Ciccio	Appulo/Valnova(f6)//(f5)Valforte/Patrizio	Italy	1996
19	Cirta	Hedba-3/Gerardo-vz-619	Algeria	2000
20	Core	Platani/Gianni	Italy	2008
21	GTA Dur	Crane/4/Polonicum PI185309//T.glutin enano/2* Tc60/3/Gll	CIMMYT	1972
22	INRAT 69	Mahmoudi/(bd-2777)Kyperounda	Tunisia	1969
23	Korifla	Durum-dwarf-s-15/Crane//Geier	ICARDA	1987
24	Mansourah	Bread wheat/MBB	Algeria	2012
25	Massinissa	Ofanto/Bousselem	Algeria	2012
26	Megress	Ofanto/Waha//MBB	Algeria	2007
27	Mexicali 75	Gerardo-vz-469/3/Jori(sib)//Nd-61-130/Leeds	CIMMYT	1975
28	Ofanto	Ademelio/Appulo	Italy	1990
29	Oued El Berd	Gta dur/Ofanto	Algeria	2013
30	Polonicum	Triticum polinicum/Zenati boulette 1953-58	France	1973
31	Sahell	Cit"s"/4/Tace/4*tc//2*zb/wls/3/aa"s"/5/Ruff"s"/Albe"s"	CYMMIT	1977
32	Simeto	Capeiti-8/Valnova	Italy	1988
33	Sitifis	Bousselam/Ofanto	Algeria	2011
34	Vitron	Turkey77/3/Jori/Anhinga//Flamingo	Spain	1987
35	Waha	Plc/Ruff//Gta's/3/ Rolette	ICARDA	1986
36	Wahbi	Bidi 17/Waha//Bidi 17	Algeria	2002
37	ZB × Fg	Zb/fg"s" lk/3/ko 120/4/Ward cs 10604	Algeria	1983

*: Backcross.

2.2. Root Phenotyping

2.2.1. Preparation of Seeds

Twelve seeds of uniform size and healthy aspects were visually selected from each genotype and surface sterilized in a sodium hypochlorite solution (1.25% + one detergent drop, Mistol Henkel Iberica®). Seeds of each genotype were soaked and shaken in the solution for 15–20 min. Then, they were rinsed four times with sterile deionized water, in sterile conditions. Twelve seeds of each genotype were placed in Petri dishes, each with two filter papers soaked with 4 mL of sterile water. Then the Petri dishes were placed in a dark room at 4 °C for four days, and then at 22 °C/18 °C in a growth chamber with a 12 h light/darkness photoperiod for about 16 h.

Finally, the pre-germinated grains were transferred to the rhizo-slide system, described in detail in the next section and Figure S1. The experiment was carried out at the Laboratory of Cellular Biology and Genetics, Department of Biomedicine and Biotechnology of the University of Alcalá, Spain.

2.2.2. The Rhizo-Slide System

The rhizo-slide system was constructed as a sandwich made with glass plate, black cardboard, filter paper, and a black plastic sheet. Sheets of A4-size black cardboard (180 g/m^2, www.liderpapel.com) and filter papers were previously sterilized in an autoclave and then soaked in the nutritive solution Aniol [40]. The nutritive solution was prepared by dissolving 0.5550 g of Ca Cl$_2$, 0.8215 g of KNO$_3$, 0.6352 g of MgCl$_2$·6H$_2$O, 0.0165 g of (NH$_4$)$_2$SO, 0.0400 g of NH$_4$NO$_3$ in 100 mL of distilled water, to which 500 μL/L of Plant Preservative Mixture (PPM^{TM}, Plant Cell Technology) at pH 5.8 was added. Each 8 mL was used to prepare 1 L of nutritive solution. Black cardboard with a nick made at the top center was placed on a glass plate with the same dimensions; then the pre-germinated grain (with embryonic part downward) was positioned just below the nick and covered by a filter paper. A black plastic sheet was used to cover the filter paper to ensure obscurity for roots, shifted ~2 cm upwards to allow better contact of the cardboard, and filter paper sheets with the nutritive solution. Two rhizo-slides were confronted to each other by the glass plate side, and the set was placed vertically in a glass box (internal dimensions of 32.2, 22, and 16 cm, length, width, and height) with two liters of the nutritive solution at the bottom, and then secured with two paper clips. Each glass box held 6 glass plates with two rhizoslides each, for a total of 12 seedlings, consisting of two genotypes, 6 seedlings for each (Figure S1). In total, each genotype was replicated 12 times. More details on the system are found in Ruiz et al. (2018) [25].

Once placed in the rhizoslides and the glass boxes, the seedlings were grown in a growth chamber for 7 days at 22/18 °C and 12/12 h photoperiod, day/night. The 37 genotypes were processed in batches of 6. Pre-germinated seeds of each 6 genotypes were placed into six glass boxes, each holding 6 seeds of two different genotypes. A complete batch comprised six boxes, three glass boxes prepared each Monday, and three each Thursday, every week. In total, 7 batches (14 runs) were performed until the experiment was completed (accounting for some seedlings that had to be replicated for various reasons). The set of genotypes for each run was selected randomly.

During the experiment, the boxes were replenished with distilled water every two days, to refill to the initial solution level. At the same time, to minimize seedling failure, each single seedling received 10 mL of the nutritive solution, applied with a pipette, near each seed. On the eighth day, the rhizo-slides were opened and shoots were immediately collected. The fresh roots were scanned using a Canon "LiDE210" scanner at 300 ppi to capture the first image then overlapped roots were manually separated and a second scan was done. The individual plant shoot dry weight (SDW) was obtained after oven-drying at 80 °C for six hours.

2.2.3. Image Analysis

The two images of a rhizo-slide were analyzed using SmartRoot software v.3.32 [41] plugin for ImageJ1.46R (http://imagej.nih.gov/ij/download.html). The first image was used to measure only root angles and the second one to assess the other root traits using manual and semi-automatic SmartRoot procedures. Each root of the seedling was traced, semi-automatically, and then SmartRoot automatically generated the corresponding traits. In total, ten variables from the Smartroot output were recorded for each seedling: total root length (TRL), primary root length (PRL), mean length of the other seminal roots (MRL), total root surface area (Surface), mean root diameter (Diameter), total root volume (Volume), root number (RN), and shoot dry weight (SDW). The root angle was determined for each root with respect to the vertical (90°). From this determination, we extracted the maximum vertical angle (MVA) represented by the root growing with the steepest angle, the least vertical angle (LVA) represented by the root growing with the widest angle, and mean vertical angle (MRA) of all the roots, for each seedling.

2.3. Statistical Analysis

The experiment was considered a completely randomized design, with 12 replicates per genotype. Statistical analyses were performed using the REML (Restricted maximum likelihood) procedure with Genstat 18 [42]. Genotypes were considered as fixed factors and replications were considered as a random factor. The "Genotype" factor (n-1 degrees of freedom) was broken down into a single degree of freedom comparison of landraces vs. cultivars (named "Type" effect), and a "within type" factor (n-2) which corresponds to the variation of genotypes within each type. Multiple means separation was carried out using LSD at 0.05 level, for variables in which the F-value for "Genotypes" was significant. A principal component analysis (PCA) and a hierarchical cluster analysis (HC) were performed using the R package FactoMineR [43]. The hcut function was used for tree cutting levels truncation. The R package Factoextra [44] was employed for extracting and visualizing the results. Broad-sense heritability (h^2) was calculated on an entry mean basis using the REML procedure, as follows:

$$h^2 = \sigma^2{}_g/(\sigma^2{}_g + {}_({\sigma^2{}_e/r)) \tag{1}$$

where $\sigma^2{}_g$ is the genotypic variance, $\sigma^2{}_e$ is the error variance and r is the number of replications.

3. Results

3.1. Genotypic Variability

We found remarkable genetic variability for all measured traits, as revealed by the highly significant differences among genotypes in the analyses of variance (Table 2). Significant differences were also found in the "type" comparison for most traits, except for Diameter and RN (Table 2). For the other traits, the mean squares for type were 4 to 12 times larger than those for genotypes.

The means of landraces showed higher or equal mean values compared to cultivars for all traits, except root angle (MRA, LVA, and MVA), which was higher in cultivars (Figure 1). It is worth mentioning that the landrace group presented higher root depth (PRL) than the cultivars.

All traits but SDW were root-related traits so, henceforward all the traits will be referred to generally as root traits unless stated otherwise. All traits (except RN) showed a near-normal distribution (Figure 1) which denotes their polygenic control. A wide range of phenotypic values was observed for most traits (Table 2). The landrace group showed a larger range of variation for TRL, Surface, Volume, and SDW than the cultivars. For the other traits, the cultivars had higher ranges of variation (Tables S1 and S2).

The coefficients of variation (CV) ranged from small values like 5.19 (Diameter) to 24.60 (MRA, Table 2). The exception was the large CV found for MVA, 59.82. When calculated separately for landraces and cultivars, slightly higher CV for most traits were found in landraces compared to cultivars (Table S2). All the traits exhibited high broad sense heritability (h^2), ranging from 0.80 for MVA to 0.98 for MRA (Table 2).

Table 2. Descriptive statistics, broad sense heritability (h^2), ANOVA summary, and correlation coefficients for the root traits assessed in 37 durum wheat cultivars.

Traits	Descriptive Statistics					ANOVA		Correlation									
	Min	Mean	Max	CV	h^2	Genotype	Type	TRL	Surface	Volume	Diameter	PRL	SDW	MRA	LVA	MVA	RN
TRL (cm)	54.28	98.49	137.22	13.51	0.90	***	***	1	***	***	ns	***	***	***	**	***	***
Surface (cm^2)	8.25	16.77	24.82	14.74	0.90	***	***	0.95	1	***	***	***	***	***	ns	***	***
Volume (cm^3)	0.0962	0.2326	0.3721	17.03	0.90	***	***	0.83	0.97	1	***	***	***	***	ns	**	***
Diameter (cm)	0.0447	0.0538	0.0620	5.19	0.87	***	ns	0.08	0.38	0.59	1	***	***	***	*	**	***
PRL (cm)	14.32	26.51	32.06	10.76	0.90	***	***	0.55	0.55	0.52	0.17	1	***	ns	**	***	ns
SDW (g)	5.70	14.72	23.50	18.66	0.94	***	***	0.55	0.63	0.64	0.40	0.58	1	ns	**	*	ns
MRA (°)	1.20	30.36	45.69	24.60	0.98	***	***	−0.39	−0.32	−0.23	0.20	−0.01	0.00	1	***	***	ns
LVA (°)	20.40	42.71	61.47	17.81	0.92	***	***	−0.28	−0.23	−0.18	0.12	0.08	0.01	0.76	1	***	ns
MVA (°)	0.00	14.10	43.49	59.82	0.80	***	***	−0.25	−0.20	−0.15	0.12	−0.27	−0.13	0.62	0.25	1	**
RN (no.)	4.00	5.32	6.00	9.13	0.87	***	ns	0.37	0.32	0.26	−0.22	−0.01	0.06	−0.09	0.02	−0.14	1

*, **, ***: sources of variation in the analyses of variance or correlation coefficients significant at $p < 0.05$, 0.01, and 0.001, respectively. TRL: Total root length, Surface: total root surface area, Volume: Total root volume, Diameter: mean root diameter, PRL: Primary root length, SDW: Shoot dry weight, MRA: Mean root angle, LVA: Least vertical angle, MVA: Maximum vertical angle, RN: Root number.

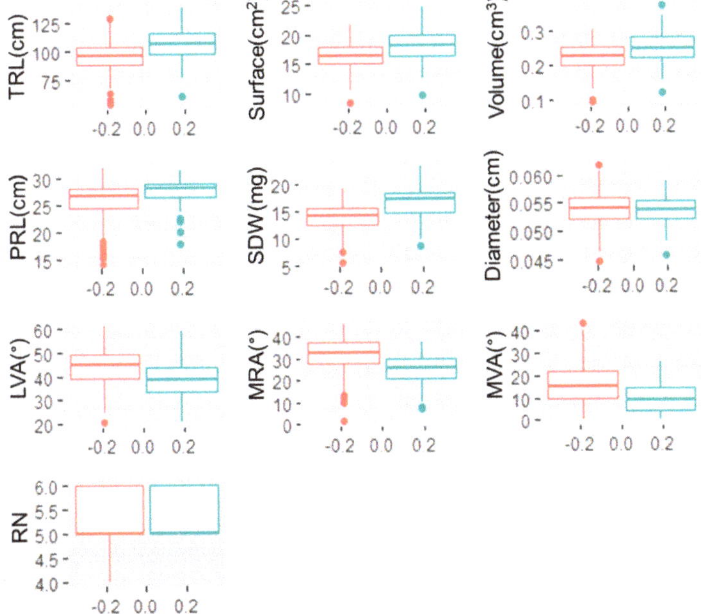

Figure 1. Boxplots for root traits for the cultivar (red) and landrace (blue) groups. Horizontal lines splitting the boxes indicate the median values; box limits indicate the 25th and 75th percentiles; whiskers extend 1.5 times the interquartile range from the 25th and 75th percentiles; outliers are represented by dots. Variable names coded as in Table 2. Genotype mean values are in Table S1.

3.2. Relationships between Traits

Highly significant correlations were found between most traits (Table 2). TRL, Surface and Volume were highly and positively correlated among them. There were moderate positive correlations between TRL, surface, and Volume, with PRL, SDW, RN, and negative ones with root angle variables (seedlings with higher TRL, Surface, and Volume tended to have steeper root angles). Seedlings with higher RN tended to have roots with thinner root diameter, indicating that there could be some kind of compensation between these traits (more roots with a finer diameter and vice versa). Interestingly, seedlings with higher primary root length produced more shoot biomass. Performing correlations between traits within each group (cultivars and landraces) showed, in general, similar patterns to the correlations performed for the entire dataset (Table S2). The moderate relationship of PRL with MVA and RN disappeared in the landrace group, compared to the cultivars and the whole dataset (Table S2).

3.3. Time Trends of Root Traits

When the genotypic means were plotted against year of release of the genotypes, different trends were observed (Figure 2 and Figure S2), in which, all the traits presented significant regression coefficients except Diameter, RN and PRL (Table S3). This trend was largely influenced by the comparison of landraces vs. cultivars because landraces are older. The trend was positive or negative depending on the trait. Overall, cultivars reduced their seminal root length and developed a shallower root angle compared to landraces (Figure 2). The root surface and volume of root presented the same trend as root length, as they were highly correlated, as mentioned above. MVA and LVA showed the same trend as MRA. No substantial variation was observed for RN, Diameter, and PRL. Regarding the shoot, a remarkable and steady reduction in SDW over the years was detected (Figure 1 and Figure S2).

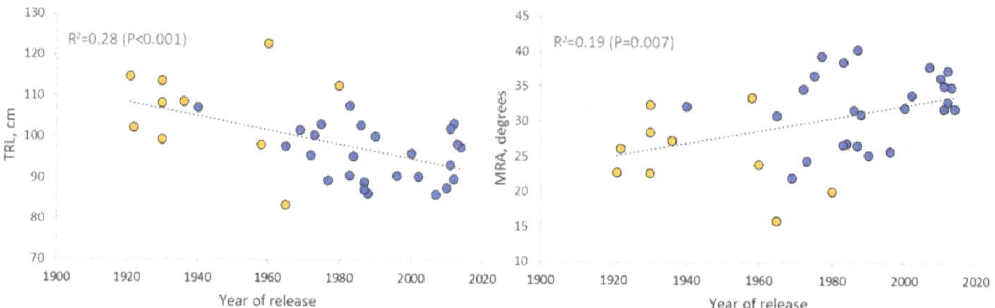

Figure 2. Time trends for total root length (TRL) and mean root angle (MRA) in seedlings of 37 durum wheat varieties. Yellow symbols correspond to landraces; blue symbols correspond to cultivars. The coefficients of determination (R^2) of the regression lines are indicated in each graph.

3.4. Grouping of Genotypes According to Root Traits

The first two principal components explained 69.63% of the total variation (Figure 3A). The first component (46%) was most related to Surface, TRL, Volume, and SDW, with the respective contributions of 20.19, 19.04, 18.16, and 11.51 (Table S4). MRA, LVA, and Diameter had the highest loadings for the second component (PC2). Correlations between these traits are discussed above (Table 2). Thus, the first axis (PC1) was related to root size traits and the second one to root architecture traits.

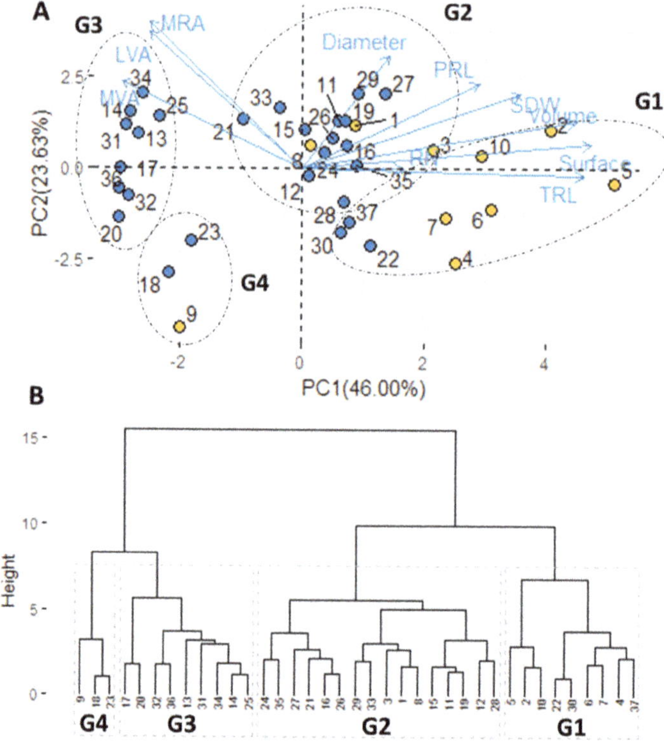

Figure 3. Biplot of the first two principal components (**A**) and dendrogram resulting from hierarchical clustering (**B**) based on seedling traits for 37 durum wheat genotypes. Ellipses in (**A**) encompass the individuals according to the clustering presented in (**B**). Yellow symbols correspond to landraces; blue symbols correspond to cultivars. Genotypes coded with numbers as in Table 1.

Genotypes were better distributed along with the first component, as a result of the contrasting position between landraces, many with large positive scores on PC1 (due to their higher root size and shoot weight) and the cultivars, with lower positive or negative scores in PC1, so the discrimination between these two groups was clear (Figure 3B). From the hierarchical classification, which was carried out based on the original data, four groups were created (G1 to G4) (Figure 3B). G1 was mostly formed by landraces. G2 was the largest one and was constituted by cultivars, and two landraces. This group was at a central position in the biplot graph (Figure 3B), presenting close to average values for most traits. G3 was located on the negative side of PC1, contrasting with G1 by having a relative smaller root size. Finally, the last group (G4) was formed by only three genotypes depicted on the negative quadrant, for both PC1 and PC2, having smaller values for both classes of root traits; fine, steeper root angle and reduced root traits related to biomass. This group included landrace Montpellier (genotype 9), which showed a special root system architecture compared to other landraces, with steeper root angle, and lower SDW, closer to two cultivars from Italy and ICARDA.

4. Discussion

The durum wheat collection used in this study was assembled to explore the seminal root variability present in a set of genotypes cultivated in Algeria, with a historical perspective on the possible changes caused by modern breeding. The method chosen enabled data acquisition and processing of 444 single plants, by one person, in two months. Its performance could be easily expanded by increasing the number of boxes and operators. Therefore, it is amenable to the scale needed for the type of studies carried out in plant genetics and breeding. Root number together with root length, the main results of this type of experiment, describe how extensively the seminal axes can potentially explore the rooting volume. These easily measurable traits at an early stage can have agronomic implications. For example, root spread angle is an additional feature whose variation can influence how crops cope with water-limited conditions and/or other environmental constraints, such as high pH, toxic ions, or low nutrient availability [45,46]. The root angular spread at an early growth stage can be used to predict the partitioning of root biomass in the soil profile at the adult plant stage [5,27,28], a feature relevant for water use efficiency in wheat [21,47]. Therefore, artificial systems are efficient at revealing phenotypic (and presumably genetic) variability, but its implications on agronomic performance must be validated later under field conditions.

4.1. Large Genotypic Variation for Seminal Root Traits

An overview of the results found in different studies sheds more light on the actual genetic variation available for seminal root traits, better than any single study. Differences among studies may be partly due to slight differences in the experimental methods, but also to the size and scope of the genetic material used. Nevertheless, some meaningful conclusions can be derived.

We found significant genetic variation for all traits. We found a range of values for the least vertical root angle (LVA) from 20.40° to 61.47°. Multiplying these values by two (range from 40.80° to 122.94°) allows the comparison of our study with others, in which the values of the total opening of the angle of the root system was reported. Our range was superior to those found by others in durum [47] and bread wheat [24,29]. Our wheat genotypes displayed similar low ranges of variation in mean root number as in similar studies in durum [47,48], with a slightly higher mean. In our genotypes, the sixth root was present in about a third of all genotypes, with no significant differences between landraces and cultivars. This is a similar proportion than found in a study of Mediterranean and North-American elite material [47], with the striking difference that in the former study they reported almost absence of the sixth seminal root in native Mediterranean materials [47]. Neither sample of landrace materials was large enough to derive definitive conclusions from these studies, but at least we can say that Algerian landraces are not more likely to lack the sixth seminal root than modern cultivars.

Based on the coefficients of variation, overall, landraces showed higher slightly variability for most traits, especially for root angle, even though the sample size was lower than for cultivars. Previous reports indicate that native Mediterranean landraces are likely to provide additional genetic variability for root architecture [46], particularly in wheat accessions that experienced long-term natural selection in drought-prone environments [49], and in barley [50]. Overall, the Algerian landraces showed sizeable genetic variation for most traits, indicating that they harbor relevant root morphology variation that should be further investigated by geneticists and breeders.

4.2. Classification of Durum Wheat Genotypes According to Root Morphology

Overall, genotypes with higher root length tended to have larger root number, as found in a previous study [47], and a narrower root angle. Other authors [24,51] found no correlation between root angle and root number. Sanguinetti et al. [47] also found no correlation of root angle with other traits and suggested that the root angle was controlled by an independent set of genes. In our study, however, given the negative correlation between MRA and root size traits, we cannot rule out that these two traits are controlled by the same set of genes.

We found that higher root length and Diameter were associated with higher SDW (r of 0.55 and 0.40, respectively, Table 2), suggesting a size effect that affected the whole plant. Rather similar observations were done in the Spanish core collection of tetraploid wheat, but the plant size effect was visible for subsp. *dicoccon* and *turgidum*, but not for *durum* [25]. Correlation between root length and volume and SDW was also found in hexaploid wheat [52]. We found no correlation between RN and MRA, in agreement with previous studies [24,51]. It seems that an overall plant size effect that affects harmonically roots and shoots is common in wheat species.

Our genotypes displayed different seminal root system patterns, from vigorous and steep to a small and shallow root system. These root patterns may be related to phylogenetic relationships, regional origin, and functional plant adaptation to different environments, as indicated in previous studies [53]. There were differences in the length of the seminal roots of single plants. This was made evident by calculating the difference between the length of the primary root (PRL), and the average of the rest (MRL). G2 and G3 had a higher difference between PRL and the mean length of other roots (MRL), compared to groups G1 and G4, which had roots with more similar lengths (Table S5). G2 genotypes combined a significantly longer primary root (Table S5) with the largest difference between it and the other seminal roots (together with G3). This rooting pattern, based on dissimilar growth of the roots, could have an impact on overall soil exploring capacity that should be explored further, particularly its usefulness in semi-arid environments, to access to stored water at deep layers at critical periods (flowering and grain filling), while keeping enough shallow roots to take advantage of in-season precipitations.

Two groups (G1 and G3) showed the highest contrast in the multivariate analysis (Figure 3). G1, with a majority of landraces, displayed a vigorous seminal root system, in contrast with G3, formed entirely by cultivars with small root systems. Our finding was in agreement with the study of a collection of 160-durum wheat landraces [33] in terms of larger seminal root size. This study found that landraces coming from the eastern Mediterranean region (Turkey), the driest and warmest areas considered in the study, showed the largest seminal root size and widest root angle compared to landraces from eastern Balkan countries. The authors claimed that these differences were due to the adaptations of landraces to the contrasting environmental conditions of these two regions. The larger root size and wider root angle from Turkish landraces would allow better exploration of the full soil profile and better water capture. Among the four groups found in this study, no one combined the highest MRA and TRL, comparable to Turkish landraces. Therefore, there could be room for improvement for the root systems of durum wheat for Algeria. Crosses to combine these traits in a single genotype should be devised, and Turkish landraces could be tested in Algerian conditions, to assess their potential.

In our germplasm, the landraces showed on an average narrower angle and higher root size. Previous studies on Mediterranean durum wheat [33] found that the genotypes with the narrowest angle came from the western Mediterranean region and that they also had heavier grains [54,55].

Additionally, it was reported that *Triticum turgidum* subsp. *dicoccon* landraces coming from cooler and wetter zones had shallower seminal root systems than those from warmer and drier areas [25]. The subsp. *durum* landraces, developed in warmer and drier areas, tended to have larger and steeper root patterns than landraces coming from cooler and wetter zones. Accordingly, the root system architecture of the Algerian landraces would indicate adaptation to a warm and dry environment. Other studies have found different root morphologies in apparent adaptation to stressful conditions. For instance, the drought-tolerant bread wheat cv. SeriM82 has a compact root system [21], associated with a limited water use early in the season, facilitating access to stored water later in the reproductive phase. Contrary to our landraces, SeriM82 exhibited less vigorous shoot growth. In contrast with our findings, a study of bread wheat germplasm grown historically in the semi-arid northwestern of China [56] found that breeding caused a narrowing of the seminal root angle, reduced root number, and increase of primary seminal root length. In that study "newer cultivars produced higher yields than older ones only at the higher sowing density, showing that increased yield results from changes in competitive behavior." This view was confirmed and expanded later [57], confirming that the advantage of new Chinese wheat cultivars came from the attenuation of inter-plant competition and increased plasticity in root morphology. A seminal root architecture with fewer, longer seminal roots with narrower root angle, would overlap less with neighbors, leading to less competition between individuals [58], and these trends agree with the hypothesis of weakening of "selfish" traits [59].

The shift in root morphology observed in Algeria in the step from landraces to modern cultivars does not conform to the scenario described in those works. There was a reduction of overall root length and volume after the advent of modern breeding, which could be consistent with the reduction of inter-plant competition but combined with the widening of the root angle, which does not bode well with that hypothesis. It seems that wheat breeding may have resulted in different trends for root morphology in different parts of the world. This could be the result of the adaptation of Algerian landraces to agronomic conditions different from current agriculture. The difference in rooting patterns between landraces from different geographical areas and cultivars may lie in the agronomic environments in which they were developed. In general, modern durum wheat cultivars were bred under high plant densities [57], whereas landraces were grown in stands with density adapted to the environment. The morphology of Algerian landraces (long seminal roots growing in steep angles) conforms to the "steep, cheap, and deep root ideotype" [60], and could be the result of adaptation to accessing water in deep soil layers. Further studies with adult plants are needed to evaluate if root features of seedlings are maintained when the competition between individuals for root growth is increased (as the seminal and nodal roots require more space and resources than just the seminal roots of the seedlings). A shovelomics experiment is being carried out with the same genotypes, which could elucidate this issue at least for some measurable traits like root angle.

The high SDW of our durum landraces compared to cultivars could be related to the lack of dwarfing genes in the landraces. This hypothesis was already confirmed previously for bread and durum wheat for some height reducing genes [61], which reduced the first seedling leaf growth in *Rht* genotypes compared with the corresponding tall wheat lines.

4.3. Conclusions and Perspectives

We have found wide genetic variability in a collection of durum wheat genotypes cultivated in Algeria and unraveled a possible historic trend that sheds light on the outcomes of modern breeding. An important issue is to what extent this variability found at the seedling stage can reflect the variability in the field with the same genetic material, more precisely, which traits can be consistent across plant phases (seedling and adult plant), enabling the selection at the early seedling stage. If this relationship is not found, then the room for the testing of seminal root traits is very limited. Experiments to evaluate this relationship are ongoing.

Overall, landraces showed a larger root size and steeper root angle. These two traits could be involved in the adaptation of landraces to water-stressed environments. The dwarfing genes seem to influence biomass partitioning; screening the current germplasm for these genes would elucidate this issue. The root size and shape in our data indicated some independence that would open opportunities to design cultivars with the desired combinations of traits.

Overall, the current genotypes present a diverse root system architecture, from compact deep-rooting to wide shallow one. This opens the opportunity to test the four different root ideotypes found (G1–G4) for functional implications under water and nutrient-limited environments. Based on the above results, we hypothesize that root architecture difference between cultivars, landraces (or steep deep vs. shallow root systems) may result in different strategies of adaptation to the availability of water and nutrients over the soil profile.

Supplementary Materials: The following are available online at http://www.mdpi.com/2073-4395/10/5/713/s1, Table S1. Mean values and standard errors (12 replications) of the seminal root traits for the 37 durum wheat genotypes. Table S2: Descriptive statistics and correlation coefficients for seedling traits, calculated separately for the cultivar and landrace groups. Table S3: Results of linear regressions of traits over years of release. Table S4: Contribution (%) of the traits to the first two principal components (PC1, PC2), as represented in Figure 3A. Table S5: Means comparison between groups formed by hierarchical clustering for the traits assessed. Figure S1. (A) One-week-old durum wheat seedlings in the rhizoslide system; (B) pictures of two genotypes contrasting for root angle, landrace Gloire de Montgolfier (left) and cultivar Oued El Berd (right); (C) schematic representation of a glass box, holding 6 glass plates, each holding two rhizoslides. Figure S2: Time trend of seminal root traits over the year of release.

Author Contributions: Conceptualization, E.I.; methodology, J.M.G.; formal analysis, R.B. and E.I.; investigation, R.B. and E.I.; resources, A.M.; software, J.M.G.; supervision, J.M.G., A.M., and E.I.; data curation, R.B.; visualization, J.M.G.; writing—original draft preparation, R.B. and J.M.G.; writing—review and editing, R.B., J.M.G., A.M., A.M.C., and E.I.; project administration, A.M.C.; funding acquisition, R.B.; A.M.; A.M.C., and E.I. All authors have read and agreed to the published version of the manuscript.

Funding: This study was funded by University of Alcalá Project CCGP2017-EXP/007. E.I. and A.M.C. acknowledge funding from the Spanish Ministry of Economy and Competitiveness and the Agencia Estatal de Investigación (Project AGL2016–80967-R), and the European Regional Development Fund.

Acknowledgments: R.B. is a recipient a PhD grant in the framework of Programme National Exceptionel (PNE 2018/2019) funded by Algerian Ministry of Higher Education and Scientific Research.

Conflicts of Interest: The authors declare no conflict of interest.

References

1. Chochois, V.; Vogel, J.P.; Rebetzke, G.J.; Watt, M. Variation in adult plant phenotypes and partitioning among seed and stem-borne roots across *Brachypodium distachyon* accessions to exploit in breeding cereals for well-watered and drought environments. *Plant Physiol.* **2015**, *168*, 953–967. [CrossRef]

2. Sinha, S.K.; Rani, M.; Kumar, A.; Kumar, S.; Venkatesh, K.; Mandal, P.K. Natural variation in root system architecture in diverse wheat genotypes grown under different nitrate conditions and root growth media. *Theor. Exp. Plant Physiol.* **2018**, *30*, 223–234. [CrossRef]

3. Araki, H.; Iijima, M. Deep rooting in winter wheat: Rooting nodes of deep roots in two cultivars with deep and shallow root systems. *Plant Prod. Sci.* **2001**, *4*, 215–219. [CrossRef] [PubMed]

4. Manske, G.G.B.; Vlek, P.L.G. Root Architecture-Wheat as a Model Plant. In *Plant Roots: The Hidden Half*, 3rd ed.; Waisel, Y., Eshel, A., Beeckman, T., Kafkafi, U., Eds.; Marcel Dekker: New York, NY, USA, 2002; pp. 249–259.

5. Maccaferri, M.; El-Feki, W.; Nazemi, G.; Salvi, S.; Canè, M.A.; Colalongo, M.C.; Stefanelli, S.; Tuberosa, R. Prioritizing quantitative trait loci for root system architecture in tetraploid wheat. *J. Exp. Bot.* **2016**, *67*, 1161–1178. [CrossRef] [PubMed]

6. Esau, K. *Plant Anatomy*, 2nd ed.; John Wiley: New York, NY, USA, 1965; 767p.

7. Lynch, J. Root architecture and plant productivity. *Plant Physiol.* **1995**, *109*, 7–13. [CrossRef] [PubMed]

8. Ahmadi, J.; Pour-Aboughadareh, A.; Fabriki-Ourang, S.; Mehrabi, A.A.; Siddique, K.H.M. Screening wheat germplasm for seedling root architectural traits under contrasting water regimes: Potential sources of variability for drought adaptation. *Arch. Agron. Soil Sci.* **2018**, *64*, 1351–1365. [CrossRef]

9. York, L.M.; Slack, S.; Bennett, M.J.; Foulkes, M.J. Wheat shovelomics I: A field phenotyping approach for characterizing the structure and function of root systems in tillering species. *BioRxiv* **2018**, 280875. [CrossRef]

10. Nguyen, V.L.; Stangoulis, J. Variation in root system architecture and morphology of two wheat genotypes is a predictor of their tolerance to phosphorus deficiency. *Acta Physiol. Plant.* **2019**, *41*, 109. [CrossRef]
11. Bengough, A.G.; Gordon, D.C.; Al-Menaie, H.; Ellis, R.P.; Allan, D.; Keith, R.; Thomas, W.T.B.; Forster, B.P. Gel observation chamber for rapid screening of root traits in cereal seedlings. *Plant Soil* **2004**, *262*, 63–70. [CrossRef]
12. Watt, M.; Moosavi, S.; Cunningham, S.C.; Kirkegaard, J.A.; Rebetzke, G.J.; Richards, R.A. A rapid, controlled-environment seedling root screen for wheat correlates well with rooting depths at vegetative, but not reproductive, stages at two field sites. *Ann. Bot.* **2013**, *112*, 447–455. [CrossRef]
13. Richard, C.A.; Hickey, L.T.; Fletcher, S.; Jennings, R.; Chenu, K.; Christopher, J.T. High-throughput phenotyping of seminal root traits in wheat. *Plant Methods* **2015**, *11*, 13. [CrossRef] [PubMed]
14. González, J.M.; Friero, E.; Selfa, L.; Froilán, S.; Jouve, N.A. Comparative study of root system architecture in seedlings of *Brachypodium* spp. using three plant growth supports. *Cereal Res. Commun.* **2016**, *44*, 69–78. [CrossRef]
15. Adeleke, E.; Millas, R.; McNeal, W.; Faris, J.; Taheri, A. Variation analysis of root system development in wheat seedlings using root phenotyping system. *Agronomy* **2020**, *10*, 206. [CrossRef]
16. Richard, C.; Christopher, J.; Chenu, K.; Borrell, A.; Christopher, M.; Hickey, L. Selection in early generations to shift allele frequency for seminal root angle in wheat. *Plant Genome* **2018**, *11*, 170071. [CrossRef] [PubMed]
17. El Hassouni, K.; Alahmad, S.; Belkadi, B.; Filali-Maltouf, A.; Hickey, L.T.; Bassi, F.M. Root system architecture and its association with yield under different water regimes in durum wheat. *Crop Sci.* **2018**, *58*, 2331–2346. [CrossRef]
18. Bai, C.; Liang, Y.; Hawkesford, M.J. Identification of QTLs associated with seedling root traits and their correlation with plant height in wheat. *J. Exp. Bot.* **2013**, *64*, 1745–1753. [CrossRef]
19. Tuberosa, R.; Salvi, S.; Sanguineti, M.C.; Landi, P.; Maccaferri, M.; Conti, S. Mapping QTLs regulating morpho-physiological traits and yield: Case studies, shortcomings and perspectives in drought-stressed maize. *Ann. Bot.* **2002**, *89*, 941–963. [CrossRef]
20. Tuberosa, R.; Sanguineti, M.C.; Landi, P.; Giuliani, M.M.; Salvi, S.; Conti, S. Identification of QTLs for root characteristics in maize grown in hydroponics and analysis of their overlap with QTLs for grain yield in the field at two water regimes. *Plant Mol. Biol.* **2002**, *48*, 697–712. [CrossRef]
21. Manschadi, A.M.; Christopher, J.; De Voil, P.; Hammer, G.L. The role of root architectural traits in adaptation of wheat to water-limited environments. *Funct. Plant Biol.* **2006**, *33*, 823–837. [CrossRef]
22. Paez-Garcia, A.; Motes, C.M.; Scheible, W.R.; Chen, R.; Blancaflor, E.B.; Monteros, M.J. Root traits and phenotyping strategies for plant improvement. *Plants* **2015**, *4*, 334–355. [CrossRef]
23. Løes, A.K.; Gahoonia, T.S. Genetic variation in specific root length in Scandinavian wheat and barley accessions. *Euphytica* **2004**, *137*, 243–249. [CrossRef]
24. Manschadi, A.M.; Hammer, G.L.; Christopher, J.T.; Devoil, P. Genotypic variation in seedling root architectural traits and implications for drought adaptation in wheat (*Triticum aestivum* L.). *Plant Soil* **2008**, *303*, 115–129. [CrossRef]
25. Ruiz, M.; Giraldo, P.; González, J.M. Phenotypic variation in root architecture traits and their relationship with eco-geographical and agronomic features in a core collection of tetraploid wheat landraces (*Triticum turgidum* L.). *Euphytica* **2018**, *214*, 54. [CrossRef]
26. Li, T.; Ma, J.; Zou, Y.; Chen, G.; Ding, P.; Zhang, H.; Yang, C.; Mu, Y.; Tang, H.; Liu, Y.; et al. Quantitative trait loci for seeding root traits and the relationships between root and agronomic traits in common wheat. *Genome* **2020**, *63*, 27–36. [CrossRef] [PubMed]
27. Alahmad, S.; El Hassouni, K.; Bassi, F.M.; Dinglasan, E.; Youssef, C.; Quarry, G.; Able, J.A.; Christopher, J.A. Major root architecture QTL responding to water limitation in durum wheat. *Front. Plant Sci.* **2019**, *10*, 436. [CrossRef]
28. Ali, M.L.; Luetchens, J.; Nascimento, J.; Shaver, T.M.; Kruger, G.R.; Lorenz, A.J. Genetic variation in seminal and nodal root angle and their association with grain yield of maize under water-stressed field conditions. *Plant Soil* **2015**, *397*, 213–225. [CrossRef]
29. Nakamoto, T.; Oyanagi, A. The direction of growth of seminal roots of *Triticum aestivum* L. and experimental modification thereof. *Ann. Bot.* **1994**, *73*, 363–367. [CrossRef]
30. Masi, C.E.A.; Maranville, J.W. Evaluation of sorghum root branching using fractals. *J. Agric. Sci.* **1998**, *131*, 259–265. [CrossRef]

31. Liao, H.; Rubio, G.; Yan, X.; Cao, A.; Brown, K.M.; Lynch, J.P. Effect of phosphorus availability on basal root shallowness in common bean. *Plant Soil* **2001**, *232*, 69–79. [CrossRef]

32. Lynch, J.P.; Brown, K.M. Topsoil foraging–an architectural adaptation of plants to low phosphorus availability. *Plant Soil* **2001**, *237*, 225–237. [CrossRef]

33. Roselló, M.; Royo, C.; Sanchez-Garcia, M.; Soriano, J.M. Genetic dissection of the seminal root system architecture in Mediterranean durum wheat landraces by genome-wide association study. *Agronomy* **2019**, *9*, 364. [CrossRef]

34. El-Beltagy, A.; Madkour, M. Impact of climate change on arid lands agriculture. *Agric. Food Secur.* **2012**, *1*, 3. [CrossRef]

35. Nazco, R.; Villegas, D.; Ammar, K.; Pena, R.J.; Moragues, M.; Royo, C. Can Mediterranean durum wheat landraces contribute to improved grain quality attributes in modern cultivars? *Euphytica* **2012**, *185*, 1–17. [CrossRef]

36. Lopes, M.S.; El-Basyoni, I.; Baenziger, P.S.; Singh, S.; Royo, C.; Ozbek, K.; Aktas, H.; Ozer, E.; Ozdemir, F.; Manickavelu, A.; et al. Exploiting genetic diversity from landraces in wheat breeding for adaptation to climate change. *J. Exp. Bot.* **2015**, *66*, 3477–3486. [CrossRef]

37. Araujo, S.S.; Beebe, S.; Crespi, M.; Delbreil, B.; González, E.M.; Gruber, V.; Lejeune-Henaut, I.; Link, W.; Monteros, M.J.; Prats, E.; et al. Abiotic stress responses in legumes: Strategies used to cope with environmental challenges. *Crit. Rev. Plant Sci.* **2015**, *34*, 237–280. [CrossRef]

38. Rao, I.M.; Beebe, S.E.; Polania, J.; Grajales, M.; Cajiao, C.; Ricaurte, J.; García, R.; Rivera, M. Evidence for genotypic differences among elite lines of common bean in the ability to remobilize photosynthate to increase yield under drought. *J. Agric. Sci.* **2017**, *155*, 857–875. [CrossRef]

39. Lynch, J.P. Steep, cheap and deep: An ideotype to optimize water and N acquisition by maize root systems. *Ann. Bot.* **2013**, *112*, 347–357. [CrossRef]

40. Aniol, A. Induction of aluminum tolerance in wheat seedlings by low doses of aluminum in the nutrient solution. *Plant Physiol.* **1984**, *75*, 551–555. [CrossRef]

41. Lobet, G.; Pagès, L.; Draye, X. A novel image-analysis toolbox enabling quantitative analysis of root system architecture. *Plant Physiol.* **2011**, *157*, 29–39. [CrossRef]

42. Payne, R.W.; Murray, D.A.; Harding, S.A.; Baird, D.B.; Soutar, D.M. *GenStat for Windows Introduction*, 12th ed.; VSN International: Hemel Hempstead, UK, 2009.

43. Le, S.; Josse, J.; Husson, F. FactoMineR: An R Package for Multivariate Analysis. *J. Stat.* **2008**, *25*, 18.

44. Kassambara, A.; Mundt, F. Package 'factoextra'. *Extr. Vis. Results Multivar. Data Anal.* **2017**, *76*. Available online: http://www.sthda.com/english/rpkgs/factoextra (accessed on 2 March 2020).

45. Devaiah, B.N.; Nagarajan, V.K.; Raghothama, K.G. Phosphate homeostasis and root development in Arabidopsis is synchronized by the zinc finger transcription factor ZAT6. *Plant Physiol.* **2007**, *145*, 147–159. [CrossRef] [PubMed]

46. Da Silva, A.; Bruno, I.P.; Franzini, V.I.; Marcante, N.C.; Benitiz, L.; Muraoka, T. Phosphorus uptake efficiency, root morphology and architecture in Brazilian wheat cultivars. *J. Radioanal. Nucl. Chem.* **2016**, *307*, 1055–1063. [CrossRef]

47. Sanguineti, M.C.; Li, S.; Maccaferri, M.; Corneti, S.; Rotondo, F.; Chiari, T.; Tuberosa, R. Genetic dissection of seminal root architecture in elite durum wheat germplasm. *Ann. Appl. Biol.* **2007**, *151*, 291–305. [CrossRef]

48. Cane, M.A.; Maccaferri, M.; Nazemi, G.; Salvi, S.; Francia, R.; Colalongo, C.; Tuberosa, R. Association mapping for root architectural traits in durum wheat seedlings as related to agronomic performance. *Mol. Breed.* **2014**, *34*, 1629–1645. [CrossRef]

49. Robertson, B.M.; Waines, J.G.; Gill, B.S. Genetic variability for seedling root number in wild and domesticated wheats. *Crop Sci.* **1979**, *19*, 843–847. [CrossRef]

50. Grando, S.; Ceccarelli, S. Seminal root morphology and coleoptile length in wild (*Hordeum vulgare* ssp. *spontaneum*) and cultivated (*Hordeum vulgare* ssp. *vulgare*) barley. *Euphytica* **1995**, *86*, 73–80.

51. O'Brien, L. Genetic variability of root growth in wheat (*Triticum aestivum* L.). *Aust. J. Agric. Res.* **1979**, *30*, 587–595. [CrossRef]

52. Narayanan, S.; Mohan, A.; Gill, K.S.; Prasad, P.V. Variability of root traits in spring wheat germplasm. *PLoS ONE* **2014**, *9*, e100317. [CrossRef]

53. Bodner, G.; Leitner, D.; Nakhforoosh, A.; Sobotik, M.; Moder, K.; Kaul, H.P. A statistical approach to root system classification. *Front. Plant Sci.* **2013**, *4*, 292. [CrossRef]

54. Royo, C.; Nazco, R.; Villegas, D. The climate of the zone of origin of Mediterranean durum wheat (*Triticum durum* Desf.) landraces affects their agronomic performance. *Genet. Resour. Crop Evol.* **2014**, *61*, 1345–1358. [CrossRef]

55. Soriano, J.M.; Villegas, D.; Aranzana, M.J.; Del Moral, L.F.G.; Royo, C. Genetic structure of modern durum wheat cultivars and Mediterranean landraces matches with their agronomic performance. *PLoS ONE* **2016**, *11*, e0160983. [CrossRef] [PubMed]

56. Zhu, Y.H.; Weiner, J.; Yu, M.X.; Li, F.M. Evolutionary agroecology: Trends in root architecture during wheat breeding. *Evol. Appl.* **2019**, *12*, 733–743. [CrossRef] [PubMed]

57. Song, L.; Zhang, D.W.; Li, F.M.; Fan, X.W.; Ma, Q.; Turner, N.C. Drought stress: Soil water availability alters the inter- and intra-cultivar competition of three spring wheat cultivars bred in different eras. *J. Agron. Crop Sci.* **2010**, *196*, 323–335. [CrossRef]

58. De Parseval, H.; Barot, S.; Gignoux, J.; Lata, J.; Raynaud, X. Modelling facilitation or competition within a root system: Importance of the overlap of root depletion and accumulation zones. *Plant Soil* **2017**, *419*, 97–111. [CrossRef]

59. Weiner, J.; Du, Y.L.; Zhang, C.; Qin, X.L.; Li, F.M. Evolutionary agroecology: Individual fitness and population yield in wheat (*Triticum aestivum*). *Ecology* **2017**, *98*, 2261–2266. [CrossRef]

60. Kembel, S.W.; Cahill, J.F. Independent evolution of leaf and root traits within and among temperate grassland plant communities. *PLoS ONE* **2011**, *6*, e19992. [CrossRef]

61. Ellis, M.H.; Rebetzke, G.J.; Chandler, P.; Bonnett, D.; Spielmeyer, W.; Richards, R.A. The effect of different height reducing genes on the early growth of wheat. *Funct. Plant Biol.* **2004**, *31*, 583–589. [CrossRef]

Article

Effects of Genotype, Growing Season and Nitrogen Level on Gluten Protein Assembly of Durum Wheat Grown under Mediterranean Conditions

Anna Gagliardi [1], Federica Carucci [1], Stefania Masci [2], Zina Flagella [1], Giuseppe Gatta [1,*] and Marcella Michela Giuliani [1,*]

[1] Department of Agricultural Food and Environmental Science, University of Foggia, 71122 Foggia FG, Italy; anna.gagliardi@unifg.it (A.G.); federica.carucci@unifg.it (F.C.); zina.flagella@unifg.it (Z.F.)
[2] Department of Agricultural and Forest Sciences, University of Tuscia, 01100 Viterbo, Italy; masci@unitus.it
*   Correspondence: giuseppe.gatta@unifg.it (G.G.); marcella.giuliani@unifg.it (M.M.G.)

Received: 24 March 2020; Accepted: 21 May 2020; Published: 25 May 2020

Abstract: Water deficit and high temperatures are the main environmental factors which affect both wheat yield and technological quality in the Mediterranean climate. The aim of the study was to evaluate the variation in the gluten protein assembly of four durum wheat genotypes in relation to growing seasons and different nitrogen levels. The genotypes, Marco Aurelio, Quadrato, Pietrafitta and Redidenari, were grown under three nitrogen levels (36, 90 and 120 kg ha^{-1}) during two growing seasons in Southern Italy. Significant lower yield and a higher protein concentration were observed in the year characterized by a higher temperature at the end of the crop cycle. The effect of the high temperatures on protein assembly was different for the genotypes in relation to their earliness. Based on PCA, in the warmer year, only the medium-early genotype Quadrato showed positive values along the "protein polymerization degree" factor, while the medium and medium-late genotypes, Marco Aurelio and Pietrafitta showed negative values along the "proteins assembly" factor. No clear separation along the two factors was observed for the early genotype Redidenari. The variation in gluten protein assembly observed in the four genotypes in relation to the growing season might help breeding programs to select genotypes suitable for facing the ongoing climate changes in Mediterranean area.

Keywords: durum wheat; glutenin polymers; gluten quality; high temperature; nitrogen fertilization

---

1. Introduction

Durum wheat (*Triticum turgidum* L., subsp. *durum* Desf.) is the most widespread cereal crop in Mediterranean countries and is grown in various climatic conditions [1].

Water deficit and high temperatures are the main environmental factors which affect both wheat yield and technological quality in the Mediterranean climate [2,3]. According to studies performed by the Intergovernmental Panel on Climate Change (IPCC), further increase in temperatures is predicted in Europe, especially in the Southern and Central parts [4,5]. In this context, the maintenance of adequate yield and quality standards is of particular interest, since the annual variability of product quality cannot be acceptable, especially for dry pasta production [6].

The wheat grain quality mainly depends on the quantity and type of gluten proteins, as well as on their aggregation/polymerization level [7,8]. In particular, gliadins, which are monomeric proteins, are mainly responsible for the viscous nature of the dough, and interact mostly via non-covalent links, while glutenin, which are polymeric proteins stabilized by disulphide bonds, determine its elasticity [9–12].

In the literature [13–22], conflicting results on the effect of high temperatures on the quality of the gluten proteins have been reported. Studies made on bread wheat suggest that when high temperatures occur in the middle of grain filling, they positively affect dough strength [13], while very high temperatures near physiological maturity can have a negative effect [14]. Ciaffi et al. [15] reported that in bread wheat, high temperatures increased the accumulation of glutenins compared to gliadins. On the contrary, O'Leary et al. [16] reported that water or thermal stress conditions throughout the grain filling period determine a delay in the synthesis of glutenins while the synthesis of gliadins is not altered. Furthermore, for common wheat, it is reported that short periods of very high temperatures can significantly reduce the proportion of SDS-insoluble polymers (UPP) [15,17], which in bread wheat (*Triticum aestivum* L.) have been positively correlated with dough viscoelasticity [7,8]. On the contrary, some authors have reported that short periods of very high temperatures can lead to an increase in the size of glutenin polymers in both soft and durum wheat [18,19]. While numerous are the studies available in the literature on the effect of high temperatures on gluten protein concentration, composition and on polymeric proteins size and distribution in common wheat [20–22], very few are the studies relative to durum wheat and to its pasta-making quality [8]. Moreover, pasta-making quality in durum wheat is mostly determined by low-molecular-weight glutenin subunits (LMW-GS), especially the B-type [23], whereas in bread wheat high molecular weight glutenin subunits (HMW-GS) play the major role in determining dough technological properties [24].

In the Mediterranean areas, after climate conditions, the nitrogen (N) availability represents the main constraint in obtaining adequate yield and quality in durum wheat [25]. Some studies on bread wheat have suggested that high doses of N tend to increase the amount of monomer proteins [26,27] and to reduce the percentage of UPP causing an increase in the extensibility of the dough [28–31]. Moreover, some authors have highlighted that the effect of nitrogen on gluten proteins composition and on polymers organization may vary according to the genotype [26,30,32]. Finally, for the same parameters, significant effect of the interaction between the high temperatures and N availability has been reported [29,33]. Malik et al. [33] highlighted that the combinations of cultivars, nitrogen and temperature were needed to explain the variation in the quantity and size distribution of the polymer proteins and their effects on the quality of the end-product. To the best of our knowledge, for durum wheat, this type of information is still lacking.

Thus, the aim of the present study was to evaluate the variation in gluten proteins quality, in terms of their capacity to assembly in a visco-elastic structure, of four durum wheat genotypes in relation to the growing season and different nitrogen levels, including a low input rate.

2. Materials and Methods

2.1. Field Trials

Four durum wheat cultivars, Marco Aurelio, Quadrato, Pietrafitta and Redidenari, that are used in an important Italian pasta supply chain, (Table 1), were grown in two rain-fed field experiments carried out at Foggia (latitude 41°46′ N and longitude 15°54′ E, 74 m a.s.l.) during two growing seasons (2016–2017 and 2017–2018, hereafter indicated as 2017 and 2018, respectively) in a clay loam soil.

Table 1. Main characteristics of the genotypes under study.

Genotype	Year of Release	Pedigree	Earliness
Pietrafitta	1999	Grazia x Isa	medium-late
Quadrato	1999	Creso x Trinakria	medium early
Marco Aurelio	2010	Orobel//Arcobaleno/Svevo	medium
Redidenari	2015	Kofa x N185	early

The main chemical and physical soil characteristics in the two experimental year, 2017 and 2018, are reported in Table 2.

Table 2. Soil physical and chemical characteristics in the two experimental years.

Soil Characteristics		2017	2018
Sand	%	21.5	25.2
Silt	%	39.8	36.2
Clay	%	38.7	38.6
pH		8.1	8.2
Organic Matter *	%	1.9	1.9
Total Nitrogen **	‰	1.3	1.3
Assimilable Phosphorus √	mg kg^{-1}	80	64
Exchangeable Potassium ◊	mg kg^{-1}	461	422
Field Capacity (−0.03 MPa)	%	37.3	33.13
Wilting Point (−1.5 MPa)	%	19.7	18.5
Bulk Density	Mgm^3	1.15	1.10

* Walkley-Black method; ** Kjeldhal method; √ Olsen method; ◊ Ammonium acetate method.

The four cultivars were sown on November 17 in 2016 and November 25 in 2017, at a seeding rate of 240 kg ha^{-1}. In both years, the experiment was in a field where the previous crop was durum wheat.

Three different nitrogen levels were adopted corresponding to 36, 90 and 120 kg ha^{-1} (N36, N90 and N120, respectively). The fertilizers used were Yara Mila Supersemina (18% nitrogen) at pre-sowing fertilization and Yara Bela Sulfan (24% nitrogen) at tillering, stem elongation and inflorescence emergence fertilization.

Each year, the experiment was arranged in a split-plot design with two factors (genotype in plots and nitrogen levels in sub-plots) and three replications; each sub-plot was 20.4 m^2.

The grain harvest was carried out at physiological maturity on 13 June 2017 and on 22 June 2018. During the experimental period, the daily climatic parameters of rainfall and temperature were recorded by a weather station near the experimental area.

2.2. Yield and Technological Quality Parameters

At harvest, grain yield (t ha^{-1}) and thousand kernel weight (TKW) were determined. Moreover, grain protein content (GPC) was performed by NIR System Infratec 1241 Analyzer (Foss, Hillerod, Denmark).

Semolina flours have been obtained from kernels milled by Bona mill 4 cylinders (sieve 180 μm).

The gluten index (GI), an indicator of the gluten strength, was determined on semolina samples using the Glutomatic system according to ICC standard 155 [34].

2.3. Calculation of %UPP and Analysis of Gluten Protein Molecular Size Distribution

The percentage of Unextractable Polymeric Proteins (%UPP) was measured trough the SE-HPLC procedure according to the method reported in Tosi et al. [35] with minor modifications. The SDS-soluble fraction was obtained by adding to the semolina a solution consisting of 0.5% (w/v) SDS in 0.05 M sodium phosphate buffer, pH 6.9 to a final concentration of 10 mg/mL (0.3 g semolina on 30 mL buffer). The mixture was stirred for 30 min at room temperature and then centrifuged at 20.000 g for 20 min at 15 °C. The supernatant was filtered through 0.45 μm PVDF filters and 20 μl were injected into a Biobasic Thermo Scientific SEC-300 Columns (300 mm × 7.8 mm; flow rate: 0.7 mL/min) and run for 40 min, with an eluent consisting of 0.05 M sodium phosphate buffer pH 6.9, containing 0.08 M NaCl and 0.1 % (w/v) SDS, using the UHPLC Ultimate 3000 Thermo scientific. Detection was at 214 nm. The SDS soluble fraction profiles were divided into four areas, corresponding to HPLC fractions F1, F2, F3 and F4 (Figure S2a). The first two areas correspond to large and medium size polymers, with both being enriched in HMW-GS (mainly F1) and B-type LMW-GS (mainly F2) of glutenin. F3 corresponds to ω-gliadins and small oligomers enriched in C-type and D-type LMW-GS subunits [23], while F4 corresponds to monomeric gliadins (α-type and β-type) and non-gluten proteins [35].

The SDS-insoluble fraction was obtained from the residue of the centrifugation step. The pellet was resuspended in 30 mL of the same extraction buffer and sonicated in a probe type sonicator (SONICS Vibracell Model VCX 130 -max output power 130 W at a frequency 20 KHz) for 30 s at 45% power setting. After centrifugation at 20.000 g for 20 min at 15 °C, the supernatant was filtered through 0.45 µm PVDF filters and 20 µL were injected into column in the same condition described above. The SDS-insoluble fraction profile (Figure S2b) showed only one peak (F1*) containing the largest glutenin polymers, insoluble in SDS solution alone, but rendered soluble by sonication.

Samples were extracted in duplicate and two replicate separations for each extraction were performed. The proportions of each peak (%F1* and%F1–%F4) were calculated as percentages of the total areas of the two chromatograms (SDS-insoluble and SDS-soluble fractions). The amount of monomeric over polymeric proteins (mon/pol) was calculated as the ratio between the sum of F3 and F4 areas and the sum of F1*, F1 and F2 areas. %UPP was determined as the ratio between F1* area and the sum of F1 and F1* areas (*100).

2.4. Statistical Analysis

The dataset was tested according to the basic assumptions of analysis of variance (ANOVA). The normal distribution of the experimental error and the common variance of the experimental error were verified through Shapiro–Wilk and Bartlett's tests, respectively. When required, Box-Cox transformations [36] were applied prior to analysis. The ANOVA procedure was performed according to a split-plot design with three replicates. Three-way ANOVA procedure was performed considering the factors (growing season, genotype and nitrogen level) as fixed factors. The statistical significance of the difference among the means was determined using Tukey's honest significance difference post hoc test at the 5% probability level. A principal component analysis (PCA) was performed on the correlation matrix of technological and SE-HPLC parameters. We obtained Principal Components (PCs) on centered and scaled variables, through diagonalization of the correlation matrix and extraction of the associated eigenvectors and eigenvalues. Grain protein content, gluten index, and SE-HPLC parameters were set as quantitative variables and used to define PCs, while genotype, N level and growing season were used as categorical variables, not considered in the computation of PCs. The coordinates of the categorical variables were calculated in order to enhance the interpretation of data and were represented as barycenter in the Principal Component biplot. The number of factors needed to adequately describe the data was determined on the basis of the eigenvalues and of the percentage of the total variance accounted by the different factors. The results of PCA were graphically represented in two-dimensional plot, using the SigmaPlot software (Systat Software, Chicago, IL, USA). ANOVA and PCA analyses were performed using the JMP software package, version 14.3 (SAS Institute Inc., Cary, NC, USA).

3. Results

3.1. Weather Condition

The climatic data related to the two growing seasons are reported in Table 3, while the rainfall distribution and maximum and minimum daily mean temperatures of the 2017 (a) and 2018 (b) crop seasons are reported in Figure S1 (Supplementary File).

The first growing season was characterized by lower rainfall compared to the second year (about 340 mm vs. 401 mm). Moreover, in the first experimental year the rain distribution was not regular, with the most intense rainfall occurred in the second decade of January, the third decade of February, the second decade of April and the first decade of May. As for the second growing season, rainfall was observed throughout the crop cycle, especially during the grain filling period, in the first ten days of May and June. In addition to rainfall, the two years differed also for the maximum temperatures during the grain filling period showing the second year the highest values. Moreover, during 2018, more days with temperatures between 30 and 35 °C and three days with temperatures higher than 35 °C, compared to 2017, occurred.

Table 3. Climatic data related to the two growing seasons.

		2017	2018
Crop cycle duration	d	209	210
Crop cycle rainfall	mm	339.9	401.4
From seeding to heading rainfall	mm	204.2	198.6
Grain filling rainfall	mm	135.7	202.5
Crop cycle Mean T	°C	12.3	13.1
Grain filling Mean T	°C	18.3	21.7
Grain filling Mean T max	°C	25.5	29.1
$30\,°C < T < 35\,°C$	d	15	23
$T > 35\,°C$	d	-	3

3.2. Yield and Technological Parameters

The analysis of variance (ANOVA) generally showed a significant effect of year (Y), genotype (G) and nitrogen (N) on the parameters considered (Table S1). The two growing seasons differently influenced the yield and the technological parameters considered. In the second growing season (Table 4), a significant lower yield, a thousand kernel weight and gluten index were observed with respect to the first one. On the contrary, grain protein content was higher in 2018 than in 2017. Relative to the nitrogen level (Table 4), a significant positive effect on grain yield was evident only under N90, while for protein content the highest value was observed under N120. Finally, the gluten index values decreased with N level increasing.

Table 4. Effect of the year, nitrogen level and genotype on grain yield, thousand kernel weight, grain protein content and gluten index.

Experimental Factors	Grain Yield (t ha^{-1})	Thousand Kernel Weight (g)	Grain Protein Content (%)	Gluten Index (-)
Year				
2017	6.66 a	60.91 a	14.53 b	64.44 a
2018	5.91 b	50.21 b	16.00 a	58.50 b
Nitrogen level				
N36	6.20 b	55.16 a	14.25 c	63.83 a
N90	6.36 a	55.90 a	15.33 b	62.71 a
N120	6.28 ab	55.62 a	16.23 a	57.88 b
Genotype				
Marco Aurelio	7.11 a	50.62 d	15.74 b	57.72 bc
Pietrafitta	5.75 c	64.47 a	15.29 c	56.50 c
Quadrato	6.42 b	54.56 b	14.08 d	61.39 b
Redidenari	5.85 c	52.60 c	15.97 a	70.28 a

For each experimental factor, values in column followed by different letters are significantly different at $P \leq 0.05$ according to Tukey's test.

Among the genotypes (Table 4), Marco Aurelio showed the highest yield value even if associated with lower thousand kernel weight. Instead, Redidenari was the genotype with the best technological quality performance showing the highest protein content and gluten index values. However, the behavior of the genotypes changed in relation to growing seasons (Table 5) and nitrogen levels adopted (Table 6). In particular, the yield decrease observed in the second year was different among the genotypes (Table 5); it was 5% and 9% for Marco Aurelio and Redidenari, and 14% and 17% for Pietrafitta and Quadrato, respectively. Moreover, Marco Aurelio in addition to presenting lower yield decrease in the second year also showed an increase in the protein content that was double compared to the other genotypes (3.1% vs. 0.4%, 1.36% and 1.07% for Pietrafitta, Quadrato and Redidenari, respectively). Finally, as for gluten index, Marco Aurelio and Redidenari showed a significant decrease in the second year, more marked for Redidenari (Table 5).

Table 5. Effect of the year x genotype interaction on grain yield, thousand kernel weight, grain protein content and gluten index.

	2017				2018			
	Marco Aurelio	Pietrafitta	Quadrato	Redidenari	Marco Aurelio	Pietrafitta	Quadrato	Redidenari
Grain yield (t ha^{-1})	7.29 a	6.20 c	7.02 ab	6.12 c	6.92 b	5.31 e	5.82 d	5.57 de
Thousand kernel weight (g)	54.82 c	71.21 a	59.25 b	58.39 b	46.42 e	57.74 b	49.88 d	46.80 e
Grain protein content (%)	14.19 f	15.11 d	13.40 g	15.43 c	17.29 a	15.47 c	14.76 e	16.50 b
Gluten index (-)	60.78 b	56.44 b	60.89 b	79.67 a	54.67 c	56.56 b	61.89 b	60.89 b

In each row, values followed by different letters are significantly different at $P \leq 0.05$ according to Tukey's test.

Table 6. Effect of the genotype x nitrogen interaction on grain yield, thousand kernel weight, grain protein content and gluten index.

		Marco Aurelio	Pietrafitta	Quadrato	Redidenari
Grain yield (t ha^{-1})					
	N36	7.16 a	5.6 f	6.46 bc	5.61 ef
	N90	7.00 a	5.99 de	6.21 cd	6.25 bcd
	N120	7.17 a	5.68 ef	6.60 b	5.68 ef
Thousand kernel weight (g)					
	N36	50.18 e	64.67 a	53.80 bcd	51.99 cde
	N90	50.76 de	65.38 a	54.22 bc	53.26 bcde
	N120	50.91 de	63.37 a	55.67 b	52.53 bcde
Grain protein content (%)					
	N36	15.33 c	14.1 e	12.8 f	14.75 d
	N90	15.60 c	15.33 c	14.03 e	16.33 b
	N120	16.28 b	16.43 ab	15.40 c	16.82 a
Gluten index (-)					
	N36	57.67 bcde	56. 00 cde	67.33 ab	74.33 a
	N90	54.33 de	61.17 bcd	66.83 abc	68.50 ab
	N120	61.17 bcd	52.33 de	50. 00 e	68. 00 ab

For each parameter, values in each row and column followed by different letters are significantly different at $P \leq 0.05$ according to Tukey's test.

The nitrogen fertilization did not significantly affect the grain yield response in Marco Aurelio, while for both Pietrafitta and Redidenari, the highest values were observed under N90 level; for Quadrato the highest value was observed under N120 even if not significantly different from N36 (Table 6). On the contrary, for all genotypes a positive effect of the nitrogen level on grain protein content was evident with the highest values observed under N120. The effect of nitrogen fertilization on gluten index was not clear; only Quadrato showed a significant decrease under N120 level (Table 6).

3.3. Measurement of %UPP and Analysis of Gluten Protein Molecular Size Distribution

SE-HPLC was used to compare the molecular size distribution of the semolina proteins by a quantitative comparison of elution profiles.

The analysis of variance performed on the percentage of SDS-insoluble protein fraction (F1*), SDS-soluble protein fraction (F1–F4), monomeric/polymeric ratio (mon/pol) and proportion of unextractable polymeric protein (%UPP) showed a general significant effect of the year (Y), genotype (G), nitrogen level (N) and their interactions (Table S2). A significant decrease of F1* and %UPP was observed in 2018 compared to 2017. Moreover, in 2018 a significant increment of the polymeric fraction, due to an increase of both F1 and F2 was observed. On the contrary, in the same year, a decrease of the monomeric fraction, due to a decrease of F4 was evident, determining also a lower mon/pol ratio with respect to 2017 (Table 7). As for the nitrogen levels, a general positive effect of N90 compared with N36 was observed for F1*, %UPP and for the monomeric fraction, while there have never been significant differences between N36 and N120 (Table 7). Finally, as for genotypes, Marco Aurelio showed higher values of %UPP and polymeric fraction, due to higher values of F1* and F2, and lower value of mon/pol ratio. On the contrary Redidenari and Pietrafitta showed lower values of polymeric fraction (again mainly due to lower F1* and F2 values) and higher values of monomeric fraction and mon/pol ratio (Table 7). Finally, Quadrato showed intermediate values for all the fraction considered. The behavior of the genotypes changed in relation to growing seasons (Table 8). A significant decrease of F1* in the second year was evident for Marco Aurelio and Pietrafitta, more marked for the former. As consequence also %UPP significantly decrease in 2018 for Marco Aurelio (13.7%) and Pietrafitta (4.2%). On the contrary, a significant increase of F1* and %UPP was observed in the second year for Quadrato. All genotypes showed the increase of F1 values in the second year and only Marco Aurelio and Pietrafitta the increase of F2 values. Also for the polymeric and monomeric fraction the effect of the growing season was observed only for Quadrato and Redidenari. In particular, in 2018 these two genotypes showed higher polymeric and lower monomeric fraction values than 2017. The increase in polymeric fraction was due mainly to the significant increase in 2018 of both F1* and F1 for Quadrato, and of F1 for Redidenari, while the decrease of the monomeric fraction was due mainly to the F4 decrease.

Table 7. Effect of the year, genotype and nitrogen level on SDS insoluble (F1*) and soluble protein fraction (F1–F4) separated by SE-HPLC, monomeric/polymeric ratio (mon/pol) and proportion of unextractable polymeric protein (%UPP).

Experimental Factors	F1*	F1	F2	F3	F4	F1*+F1	Polymeric Fraction (F1*+F1+F2)	Monomeric Fraction (F3+F4)	%UPP	mon/pol
					(%)					(-)
Year										
2017	10.66 a	24.16 b	11.52 b	22.99 a	30.68 a	34.82 b	46.34 b	53.66 a	30.20 a	1.17 a
2018	9.71 b	26.99 a	12.07 a	23.24 a	27.99 b	36.70 a	48.77 a	51.23 b	26.26 b	1.06 b
Nitrogen level										
N36	10.07 b	26.22 a	11.75 a	22.59 a	29.38 a	36.29 a	48.04 a	51.96 b	27.66 b	1.09 b
N90	10.68 a	24.69 b	11.76 a	23.46 a	29.42 a	35.37 b	47.13 b	52.87 a	30.04 a	1.13 a
N120	9.80 b	25.81 a	11.88 a	23.30 a	29.21 a	35.61 ab	47.49 ab	52.51 ab	27.00 b	1.12 ab
Genotype										
Marco Aurelio	12.32 a	25.71 a	13.22 a	22.73 b	26.02 d	38.04 a	51.26 a	48.74 c	32.18 a	0.96 c
Pietrafitta	8.46 c	25.82 a	10.91 c	24.31 a	30.50 b	34.28 c	45.20 c	54.80 a	24.64 c	1.21 a
Quadrato	10.28 b	25.38 a	11.92 b	22.97 ab	29.45 c	35.66 b	47.58 b	52.42 b	28.69 b	1.11 b
Redidenari	9.68 b	25.37 a	11.12 c	22.46 b	31.37 a	35.05 bc	46.17 c	53.83 a	27.43 b	1.17 a

For each experimental factors, values in column followed by different letters are significantly different at $P \leq 0.05$ according to Tukey's test.

Table 8. Effect of the year x genotype interaction on SDS insoluble (F1*) and soluble protein fraction (F1–F4) separated by SE-HPLC, monomeric/polymeric ratio (mon/pol) and proportion of unextractable polymeric protein (%UPP).

(%)	2017				2018			
	Marco Aurelio	Pietrafitta	Quadrato	Redidenari	Marco Aurelio	Pietrafitta	Quadrato	Redidenari
F1*	14.95 a	9.18 c	8.93 cd	9.56 c	9.70 c	7.74 d	11.62 b	9.80 c
F1	23.11 d	24.98 c	24.00 cd	24.53 c	28.31 a	26.67 b	26.76 b	26.21 b
F2	12.76 b	10.32 f	12.09 c	10.90 ef	13.68 a	11.50 cde	11.76 cd	11.34 de
F3	22.50 b	23.72 ab	23.27 ab	22.45 b	22.96 ab	24.89 a	22.66 ab	22.47 b
F4	26.68 d	31.79 a	31.70 a	32.56 a	25.36 e	29.21 c	27.19 d	30.19 b
F1*+F1	38.1 a	34.2 c	32.3 c	34.1 c	38 a	34.4 bc	38.4 a	36 b
Polymeric fraction (F1*+F1+F2)	50.83 a	44.49 c	45.03 c	45.00 c	51.68 a	45.9 bc	50.14 a	47.34 b
Monomeric fraction (F3+F4)	49.17 c	55.51 a	54.97 a	55.00 a	48.32 c	54.1 ab	49.86 c	52.66 b
UPP	39.0 a	26.7 cd	27.1 cd	28.0 bc	25.3 d	22.5 e	30.3 b	26.9 cd
mon/pol (-)	0.97 c	1.25 a	1.22 a	1.22 a	0.94 c	1.18 ab	0.99 c	1.12 b

Values in each row followed by different letters are significantly different at $P \leq 0.05$ according to Tukey's test.

Relative to the effect of the genotype x nitrogen level interaction (Table 9), a significant effect of nitrogen level on F1* was evident for Marco Aurelio and Redidenari; in particular, for the former the F1* values increased with N level increasing, while for Redidenari the highest value was observed under N90. Both of these genotypes showed also highest %UPP values under N90. Moreover, only Redidenari showed a significant effect of the nitrogen level on the polymeric and the monomeric fraction, showing under N120 lower polymeric and higher monomeric fraction values.

Table 9. Effect of the genotype x nitrogen level interaction on SDS insoluble (F1*) and soluble protein fraction (F1–F4) separated by SE-HPLC, monomeric/polymeric ratio (mon/pol) and proportion of unextractable polymeric protein (%UPP).

(%)		Marco Aurelio	Pietrafitta	Quadrato	Redidenari
F1*	N36	11.18 bcd	9.25 efg	10.26 cde	9.58 def
	N90	12.70 ab	8.33 fg	9.80 def	11.91 abc
	N120	13.10 a	7.80 g	10.77 cde	7.55 g
F1	N36	25.86 ab	26.51 ab	25.81 ab	26.72 a
	N90	25.07 b	25.10 b	25.17 b	23.42 c
	N120	26.21 ab	25.86 ab	25.18 b	25.98 ab
F2	N36	13.01 a	10.73 de	11.71 bc	11.54 bcd
	N90	13.45 a	11.02 cde	12.05 b	10.50 e
	N120	13.20 a	10.98 cde	12.02 b	11.32 bcde
F3	N36	23.10 abc	23.38 abc	22.97 abc	20.9 c
	N90	23.28 abc	24.10 ab	23.36 abc	23.09 abc
	N120	21.8 bc	25.45 a	22.57 abc	23.39 abc
F4	N36	26.85 e	30.13 bcd	29.26 d	31.26 ab
	N90	25.51 f	31.45 a	29.63 d	31.09 abc
	N120	25.70 ef	29.91 cd	29.46 d	31.77 a
F1*+F1	N36	37.04 abc	35.76 bcde	36.07 bc	36.29 bc
	N90	37.76 ab	33.43 f	34.96 cdef	35.32 cdef
	N120	39.31 a	33.66 def	35.95 bcd	33.53 ef
Polymeric fraction (F1*+F1+F2)	N36	50.04 ab	46.49 cd	47.77 bc	47.88 bc
	N90	51.22 a	44.46 d	47.01 cd	45.82 cd
	N120	52.51 a	44.64 d	47.97 bc	44.84 d
Monomeric fraction (F3+F4)	N36	49.95 cd	53.51 ab	52.23 bc	52.17 bc
	N90	48.78 d	55.54 a	52.99 ab	54.17 ab
	N120	47.49 d	55.36 a	52.03 bc	55.15 a
UPP	N36	30.2 bc	25.8 def	28.5 cd	26.2 def
	N90	33.8 a	24.9 efg	27.7 cde	33.7 a
	N120	32.5 ab	23.2 fg	29.9 bc	22.4 g
mon/pol (-)	N36	1.00 fg	1.15 b-e	1.10 def	1.10 de
	N90	0.96 g	1.25 a	1.14 cde	1.19 abcd
	N120	0.91 g	1.24 ab	1.09 ef	1.23 abc

For each parameter, values in each row and column followed by different letters are significantly different at $P \leq 0.05$ according to Tukey's test.

3.4. PCA Analysis

A principal component analysis (PCA) was performed on the correlation matrix. The results of PCA allowed two factors to be identified explaining 51% and 20.9% of total variance, respectively (Table 10). The first factor (PC1) was highly and positively associated with the largest insoluble polymers (F1*), the medium size soluble polymers (F2), the largest glutenin polymers (both insoluble and soluble; F1*+F1) and with the polymeric fraction (F1*+F1+F2). Moreover, it was highly and negatively related with the small oligomers fraction (F3), the monomeric gliadin fraction (F4), the total monomeric fraction (F3+F4) and mon/pol ratio. Thus, PC1 could be considered a factor linked to the degree of polymerization, mostly depending on the capacity to form covalent bonds. The second factor (PC2) was positively associated with gluten index (depending on the interactions among gluten proteins, both gliadins and glutenins), with the largest insoluble polymers (F1*) and with %UPP (depending on glutenin polymers size and amount) and negatively related with grain protein content (that can affect mostly gliadin accumulation) and the large size soluble polymers (F1) (that affect negatively %UPP). Thus, PC2 could be considered as a "gluten proteins assembly" factor, including the different interactions occurring in the gluten network. Both the factors linked to the degree of polymerization and the gluten proteins aggregation are major determinants of technological quality.

Table 10. Loading matrix values for the first two principal components (PC1 and PC2), considering the original variables. The corresponding percentages of accounted variation are also reported.

Original Variables	Loading Matrix Values	
	PC1	PC2
Grain protein content	0.09	−0.57
Gluten index	−0.21	0.47
F1* (%)	0.72	0.64
F1 (%)	0.29	−0.80
F2 (%)	0.54	−0.35
F3 (%)	−0.57	−0.26
F4 (%)	−0.81	0.29
F1*+F1 (%)	0.94	0.04
F1*+F1+F2 (%)	0.99	−0.09
F3+F4 (%)	−0.99	0.09
UPP (%)	0.55	0.77
mon/pol	−0.99	0.10
Percentage explained variation	**51**	**20.9**
Percentage cumulative variation	**71.9**	

In Figure 1, the biplot relative to the principal component analysis is reported. Based on the barycenter of the categorical variables (Figure 1, yellow marks), the nitrogen level did not show a clear separation along the two factors considered. On the contrary, the separation between the two years was observed mainly along the "gluten proteins assembly" factor (PC2) with the 2018 in the lower part. However, the separation between the crop seasons has to be interpreted also considering the genotype behaviors. Only for Quadrato the two years were separated mainly along the PC1 (polymerization degree factor), with the 2018 showing the positive and higher values. No clear separation was observed for the early maturing genotype Redidenari along the two PC factors. On the other hand, Marco Aurelio and Pietrafitta showed a clear separation of the two years only along the PC2, more marked for Marco Aurelio, with the 2018 showing the lower values. Finally, only the two genotypes, Marco Aurelio and Pietrafitta were clearly separated along PC1, presenting Marco Aurelio positive values and RDD negative values.

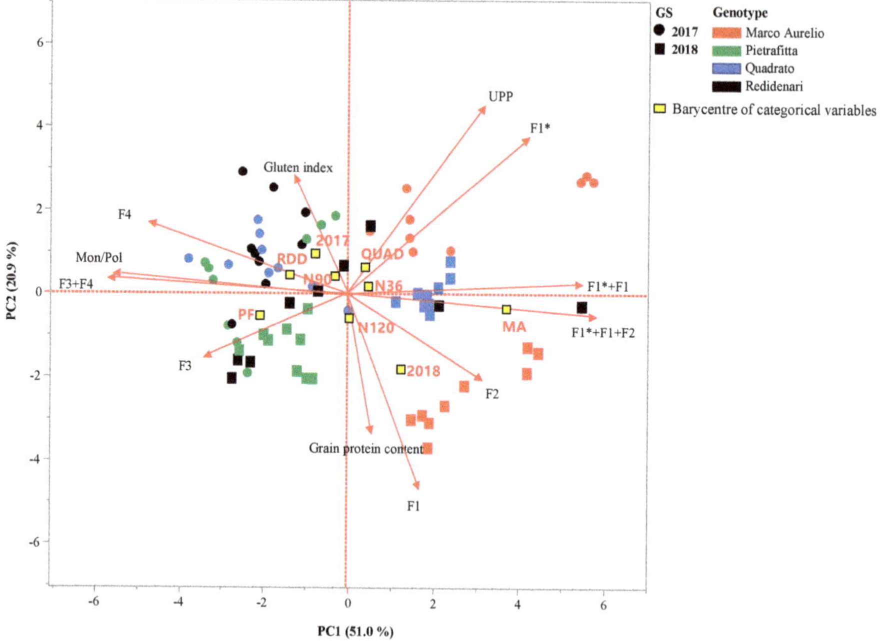

Figure 1. Biplot relative to the principal component analysis performed on grain protein content, gluten index, SDS insoluble (F1*) and soluble protein fraction (F1–F4) separated by SE-HPLC, monomeric/polymeric ratio (mon/pol) and proportion of unextractable polymeric protein. In yellow, the barycenter of the categorical variables, growing season (2017 and 2018), genotype (MA, Marco Aurelio; PF, Pietrafitta; QUAD, Quadrato; RDD, Redidenari) and nitrogen level (N36, N90 and N120) are shown.

4. Discussion

In the Mediterranean climate, the rainfall variability together with the frequency of high temperature during the grain filling period, may cause large fluctuations in durum wheat grain yield and technological quality aspects [3,37]. In semi-arid regions, a further increase in temperatures together with reduced rainfall are expected following the ongoing climate change [38,39]. This trend will influence also the crop responses to nitrogen fertilization, which depend on rainfall amount and distribution during the crop cycle, to the amount and timing of nitrogen applications as well as to the initial soil nitrogen levels [40,41]. Moreover, Malik et al. [33] highlighted that the combinations of cultivars, nitrogen and temperature are needed to explain the variation in the quantity and size distribution of the polymer proteins and their effect on the quality of the end-product. To the best of our knowledge, for durum wheat, this type of information is still lacking. The results obtained in this study represent a tile of the complex mosaic depicting the interactions among environment, fertilization and genotype.

Glutenin polymers are among the major determinants of wheat quality. Polymers are formed by different types of subunits that are functionally divided into chain terminators, chain extenders, and chain branches, according to their possibility to form one, two, or three (or more) intermolecular bonds, respectively (reviewed in [23]). The combination of these three functional glutenin classes gives rise to a range of glutenin polymers with different sizes and structures, that contributes to dough rheological properties. In general, the higher the size and amount of glutenin polymers, the better dough strength, that can be predicted by the %UPP value [7].

In our experimental condition, the two growing seasons showed a different climatic trend in terms of rainfall distribution and temperatures. Significant lower yield and thousand kernel weight, together with higher protein concentration were observed for all the genotypes in 2018, characterized by higher temperatures during the grain filling with respect to the first growing season. Moderate high temperature during grain filling, between 25 °C and 35 °C, and short periods of very high temperature (>35 °C) at the end of grain filling phase, as those we observed in the second growing season, are frequently associated with a decrease in grain yield and an increase in grain protein concentration [8,42]. However, the genotypes Marco Aurelio and Redidenari (released in 2010 and 2015, respectively) were less influenced by the growing season with respect to Quadrato and Pietrafitta (both released in 1999). The positive effect of nitrogen fertilization was clearer for the protein content than for grain yield as also reported in literature under Mediterranean climate [43–46]. However, the high yield response observed for Redidenari under N90 level was particularly interesting, indicating the possibility of limiting nitrogen inputs by adopting genotypes capable to optimize the use of nitrogen.

The growing season differently affected the gluten index, an indicator of gluten strength for durum wheat, in relation to the genotypes, showing only Marco Aurelio and Redidenari lower values in the warmer year. In bread and soft wheat, dough strength has been often positively correlated with the proportion of UPP [15,18,47–49]. As for durum wheat, the relation between %UPP and gluten index has been less investigated. In our experimental condition, this relation was genotype dependent, since only Marco Aurelio and Redidenari showed simultaneous decrease of gluten index and %UPP in the second year.

The composition and functionality of storage proteins have been significantly affected by growing season and genotype, while the effect of N fertilization level was rather small (Table S2) [50] as also resulted by PCA analysis. Several studies reported an increase in the proportions of the monomeric gliadins with increasing N availability [26,27]. In our experimental conditions, this was true only for the genotype Redidenari due to an increase of F4 component represented mainly by α/β type gliadin. An interesting result was the increase of %UPP for both Marco Aurelio and Redidenari under N90 level due to the increase of the F1*. The significant decrease of the larger insoluble polymers fraction (F1*) and %UPP observed in the second growing season for Marco Aurelio and Pietrafitta has to be discussed in relation to their earliness. Indeed, the very high temperature recorded at the end of the crop cycle (3 days with T > 35 °C) could have negatively influenced these two genotypes that are medium and medium-late maturing genotypes. This result is probably due to the fact that the assembly of the storage proteins takes place at the end of the grain filling phase [10,51,52]. Shewry et al. [53] proposed that at the end of the cycle, the loss of water favors the polymer chains contact inducing the assembly through disulphide crosslinking or through inter-chain hydrogen bonding. The effect of the temperatures on gluten protein assembly, have been studied mostly in bread wheat and only few studies are available for durum wheat. In common wheat, several research studies suggested that moderate high temperature or few days of very high temperature resulted in a significant reduction in the proportion of the SDS-insoluble protein fraction [15,17,47]. Other studies showed that the size of the glutenin polymers increased in response to short periods of very high temperature [18]. Ferreira et al. [8], in durum wheat, reported also a positive effect of the high temperature during the whole grain filling period on gluten protein assembly. Thus, the relationship between the gluten protein assembly and high temperatures is still not clear and needs more investigation. In our experimental conditions, in the second growing season, the two late maturing genotypes (Marco Aurelio and Pietrafitta), together with the decrease in F1* and %UPP showed an increase of both F2 and F1 fraction, the latter together with the other genotypes, confirming that the synthesis of the SDS soluble polymers continued also under high temperature condition [14,47]. Due to the concurrent decrease in F1* and increase in F1 and F2 fractions, Marco Aurelio and Pietrafitta did not significantly change their polymeric fraction between the two years. The increase of both %UPP and polymeric fraction observed in Quadrato and only of polymeric fraction observed in Redidenari in the second growing seasons is also linked to their earliness. Indeed, it seems like that on these genotypes, which are medium-early and early maturing, respectively, only the moderately high temperatures occurring during the grain filling acted, but not the extreme ones recorded at the end of the crop cycle. Indeed, also the results of the PCA highlighted the negative effect of the extreme temperatures on the gluten proteins assembly properties (PC2) only for Marco Aurelio and Pietrafitta, while for Quadrato a separation of the values only along the polymerization degree factor (PC1) was observed, with the warmer year showing the positive and higher values.

Because %UPP depends on protein distributions among the four areas typically used for its calculation, with the chain branchers and extenders mostly present in the fractions F1 (in particular F1*) and F2, it is important not only to select durum wheat varieties with proper glutenin compositions able to give rise to polymers of adequate size and amounts, but also that are synthetized in periods less susceptible to environmental changes, such it has occurred here for the medium early and early maturing varieties.

5. Conclusions

In the two growing seasons, the four durum wheat genotypes showed different capacities of the gluten proteins to assembly in a visco-elastic structure in relation to their earliness. In particular, in the second warmer year the late maturing genotype, Marco Aurelio and Pietrafitta showed a significant decrease of larger insoluble polymers fraction (F1*) and %UPP with a negative effect on their protein assembly level, despite Marco Aurelio always showed higher degree of polymerization. On the contrary, the medium-early and early maturing genotypes Quadrato and Redidenari, probably due to their earliness, did not change their "protein assembly level" in relation to the growing season.

The effect of N fertilization on the gluten protein polymerization and assembly was rather small, but among the N levels utilized the increase of F1*, %UPP and monomeric fraction under N90 was observed. Moreover, also the highest yield and gluten index values were obtained under N90. This was true especially for Redidenari.

In general, the effect of the growing season on the parameters evaluated was more evident than those of genotype and nitrogen level.

The results obtained in this study regarding four durum wheat genotypes clearly indicate different patterns of protein assembly in relation to the growing season, a factor that has a great influence on quality characteristics, thus contributing to the rational selection of the durum wheat genotypes, in particular those to include in supply chains, suitable for facing the ongoing climate changes in Mediterranean area.

Supplementary Materials: The following are available online at http://www.mdpi.com/2073-4395/10/5/755/s1: Table S1. Mean square of effects (year, Y; genotype, G; nitrogen level, N) resulting from analysis of variance (ANOVA) performed on yield and technological parameters. Table S2: Mean square of effects (year, Y; genotype, G; nitrogen level, N) resulting from analysis of variance (ANOVA) performed on sonicated protein fraction (F1*) and SDS-soluble protein fraction (F1–F4) separated by SE-HPLC, monomeric/polymeric ratio (mon/pol) and proportion of unextractable polymeric protein (UPP). Figure S1. Rainfall distribution and maximum and minimum mean temperatures for the two growing seasons 2017 (a) and 2018 (b). Figure S2. SE-HPLC chromatograms of SDS-extractable protein fraction (a) and of SDS-unextractable protein fraction (b).

Author Contributions: Conceptualization, M.M.G. Methodology, M.M.G. and G.G. Validation, M.M.G., G.G., S.M., Z.F. Formal analysis, A.G., G.G. and M.M.G.. Investigation, A.G., F.C. Writing—original draft preparation, A.G., F.C. Writing—review and editing, M.M.G., G.G., S.M. and Z.F. Visualization, A.G. and F.C. Supervision, M.M.G., G.G. Project administration, M.M.G. and G.G. All authors have read and agreed to the published version of the manuscript.

Funding: This research received no external funding.

Acknowledgments: We would like to show our gratitude to Luigi Toriaco (Syngenta Italia) and to Damiana Tozzi for their skilful technical assistance during the experimental trials and in the SE-HPLC analysis, respectively.

Conflicts of Interest: The authors declare no conflict of interest.

References

1. Ayadi, S.; Karmous, C.; Hammami, Z.; Trifa, Y.; Rezgui, S. Variation of durum wheat yield and nitrogen use efficiency under Mediterranean rainfed environment. *Int. J. Agric. Crop Sci.* **2016**, *7*, 693–699. [CrossRef]
2. Fagnano, M.; Fiorentino, N.; D'Egidio, M.G.; Quaranta, F.; Ritieni, A.; Ferracane, R.; Raimondi, G. Durum wheat in conventional and organic farming: Yield amount and pasta quality in southern Italy. *Sci. World J.* **2012**. [CrossRef] [PubMed]
3. Ercoli, L.; Lulli, L.; Mariotti, M.; Masoni, A.; Arduini, I. Post-anthesis dry matter and nitrogen supply and soil water availability. *Eur. J. Agron.* **2008**, *28*, 138–147. [CrossRef]
4. ICCP, Climatic Change 2011. IPCC Special Report on Renewable Energy Sources and Climate Change Mitigation. Summary for Policymakers. In Proceedings of the 11th Session of Working Group III of the IPCC, Abu Dhabi, UAE, 5–8 May 2011.
5. Bita, C.; Gerats, T. Plant tolerance to high temperature in a changing environment: Scientific fundamentals and production of heat stress-tolerant crops. *Front. Plant Sci.* **2013**, *4*, 273. [CrossRef] [PubMed]
6. Marchylo, B.A.; Dexter, J.E.; Clarke, F.R.; Clarke, J.M.; Preston, K.R. Relationships among bread-making quality, gluten strength, physical dough properties, and pasta cooking quality for some Canadian durum wheat genotypes. *Can. J. Plant Sci.* **2011**, *81*, 611–620. [CrossRef]
7. Gupta, R.B.; Popineau, Y.; Lefebvre, J.; Cornec, M.; Lawrence, G.J.; MacRitchie, F. Biochemical basis of flour properties in bread wheats. II. Changes in polymeric protein formation and dough/gluten properties associated with the loss of low Mr or high Mr glutenin subunits. *J. Cereal Sci.* **1995**, *21*, 103–116. [CrossRef]
8. Ferreira, M.S.L.; Martre, P.; Mangavel, C.; Girousse, C.; Rosa, N.N.; Samson, M.F.; Morel, M.H. Physicochemical control of durum wheat grain filling and glutenin polymer assembly under different temperature regimes. *J. Cereal Sci.* **2012**, *56*, 58–66. [CrossRef]
9. Gras, P.W.; Anderssen, R.S.; Keentok, M.; Békés, F.; Appels, R. Gluten protein functionality in wheat flour processing: A review. *Aust. J. Agric. Res.* **2001**, *52*, 1311–1323. [CrossRef]
10. Shewry, P.; Halford, N.G. Cereal seed storage proteins: Structure, properties and role in grain utilization. *J. Exp. Bot.* **2002**, *53*, 947–958. [CrossRef]
11. Triboï, E.; Martre, P.; Triboi-Blondel, A. Environmentally-induced changes in protein composition in developing grains of wheat are related to changes in total protein content. *J. Exp. Bot.* **2003**, *54*, 1731–1742. [CrossRef]
12. Anjum, F.M.; Khan, M.R.; Din, A.; Saeed, M.; Pasha, I.; Arshad, M.U. Wheat gluten: High molecular weight glutenin subunits-structure, genetics, and relation to dough elasticity. *J. Food Sci.* **2007**, *72*, 56–63. [CrossRef] [PubMed]
13. Stone, P.J.; Gras, P.W.; Nicolas, M.E. The influence of recovery temperature on the effects of a brief heat shock on wheat. III. Grain protein composition and dough properties. *J. Cereal Sci.* **1997**, *25*, 129–141. [CrossRef]
14. Stone, P.J.; Nicolas, M.E. Effect of timing of heat stress during grain filling on two wheat varieties differing in heat tolerance. II. Fractional protein accumulation. *Aust. J. Plant Physiol.* **1996**, *23*, 739–749. [CrossRef]
15. Ciaffi, M.; Tozzi, L.; Borghi, B.; Corbellini, M.; Lafiandra, D. Effect of heat shock during grain filling on the gluten protein composition of bread wheat. *J. Cereal Sci.* **1996**, *24*, 91–100. [CrossRef]
16. O'Leary, G.; Walker, C.; Panozzo, J.; Thayalakumaran, T.; McCaskill, M.; Nuttall, J.; Barlow, K.; Christy, B.; Asseng, S. Linking wheat grain quality to environmental processes. In Proceedings of the 2019 Agronomy Australia Conference, Wagga Wagga, Australia, 25–29 August 2019. Available online: www.agronomyaustralia.org/conference-proceedings (accessed on 24 February 2020).
17. Corbellini, M.; Canevar, M.G.; Mazza, L.; Ciaffi, M.; Lafiandra, D.; Borghi, B. Effect of the duration and intensity of heat shock during grain filling on dry matter and protein accumulation, technological quality and protein composition in bread and durum wheat. *Aust. J. Plant Physiol.* **1997**, *24*, 245–260. [CrossRef]
18. Don, C.; Lookhart, G.; Naeem, H.; MacRitchie, F.; Hamer, R.J. Heat stress and genotype affect the glutenin particles of the glutenin macropolymer-gel fraction. *J. Cereal Sci.* **2005**, *42*, 69–80. [CrossRef]
19. Flagella, Z.; Giuliani, M.M.; Giuzio, L.; Volpi, C.; Masci, S. Influence of water deficit on durum wheat storage protein composition and technological quality. *Eur. J. Agron.* **2010**, *33*, 197–207. [CrossRef]
20. Edwards, N.M.; Gianibelli, M.C.; McCaig, T.N.; Clarke, J.M.; Ames, N.P.; Larroque, O.R.; Dexter, J.E. Relationships between dough strength, polymeric protein quantity and composition for diverse durum wheat genotypes. *J. Cereal Sci.* **2007**, *45*, 140–149. [CrossRef]

21. Ohm, J.B.; Hareland, G.; Simsek, S.; Seabourn, B. Size-exclusion HPLC of protein using a narrow-bore column for evaluation of breadmaking quality of hard spring wheat flours. *Cereal Chem.* **2009**, *86*, 463–469. [CrossRef]

22. Malalgoda, M.; Ohm, J.B.; Meinhardt, S.; Simsek, S. Association between gluten protein composition and breadmaking quality characteristics in historical and modern spring wheat. *Cereal Chem.* **2018**, *95*, 226–238. [CrossRef]

23. D'Ovidio, R.; Masci, S. The low-molecular-weight glutenin subunits of wheat gluten. *J. Cereal Sci.* **2004**, *39*, 321–339. [CrossRef]

24. Shewry, P.R.; Gilbert, S.M.; Savage, A.W.J.; Tatham, S.; Wan, Y.F.; Belton, P.S.; Wellner, N.; D'Ovidio, R.; Bekes, F.; Halford, N.G. Sequence and properties of HMW subunit 1Bx20 from pasta wheat (*Triticum durum*) which is associated with poor end use properties. *Theor. Appl. Genet.* **2003**, *106*, 744–750. [CrossRef] [PubMed]

25. Ryan, J.; Masri, S.; Pala, M.; Singh, M. Nutrient Dynamics in a Long-Term Cereal-Based Rotation Trial in a Mediterranean Environment: Nitrogen Forms. *Commun. Soil Sci. Plant* **2009**, *40*, 931–946. [CrossRef]

26. Zhu, J.; Khan, K. Effects of genotype and environment on glutenin polymers and breadmaking quality. *Cereal Chem.* **2001**, *78*, 125–130. [CrossRef]

27. Kindred, D.R.; Verhoeven, T.M.O.; Weightman, R.M.; Swanston, J.S.; Agu, R.C.; Brosnan, J.M.; Sylvester-Bradley, R. Effects of variety and fertiliser nitrogen on alcohol yield, grain yield, starch and protein content, and protein composition of winter wheat. *J. Cereal Sci.* **2008**, *48*, 46–57. [CrossRef]

28. Johansson, E.; Prieto-Linde, M.L.; Svensson, G. Influence of nitrogen applicationrate and timing on grain protein composition and gluten strength in Swedishwheat cultivars. *J. Plant Nutr. Soil Sci.* **2004**, *167*, 345–350. [CrossRef]

29. Johansson, E.; Kuktaite, R.; Andersson, A.; Prieto-Linde, M.L. Protein polymer build-up during wheat grain development: Influences of temperature andnitrogen timing. *J. Sci. Food Agric.* **2005**, *85*, 473–479. [CrossRef]

30. Godfrey, D.; Hawkesford, M.; Powers, S.; Millar, S.; Shewry, P.R. Effects of Crop Nutrition on Wheat Grain Composition and End Use Quality. *J. Agric. Food Chem.* **2010**, *58*, 3012–3021. [CrossRef]

31. Ferrise, R.; Bindi, M.; Martre, P. Grain filling duration and glutenin polymerization under variable nitrogen supply and environmental conditions for durum wheat. *Field Crops Res.* **2015**, *171*, 23–31. [CrossRef]

32. Peckanek, U.; Karger, A.; Groger, S.; Charvat, B.; Schoggl, G.; Lelley, T. Effect of nitrogen fertilization on quantity of flour protein components, dough properties and breadmaking quality of wheat. *Cereal Chem.* **1997**, *74*, 800–805. [CrossRef]

33. Malik, A.H.; Prieto-Linde, M.L.; Kuktaite, R.; Andersson, A.; Johansson, E. Individual and interactive effects of cultivar maturation time, nitrogen regime and temperature level on accumulation of wheat grain proteins. *J. Sci. Food Agric.* **2011**, *91*, 2192–2200. [CrossRef] [PubMed]

34. International Association of Cereal Chemistry (ICC). *Standard Methods of the ICC*; ICC: Vienna, Austria, 1986.

35. Tosi, P.; Masci, S.; Giovangrossi, A.; D'Ovidio, R.; Bekes, F.; Larroque, O.; Napier, J.; Shewry, P. Modification of the low molecular weight (LMW) glutenin composition of transgenic durum wheat: Effects on glutenin polymer size and gluten functionality. *Mol Breed.* **2005**, *16*, 113–126. [CrossRef]

36. Box, G.; Cox, D. An analysis of transformations. *J. R. Stati. Soc. Ser. B Methodol.* **1964**, *26*, 211–252. [CrossRef]

37. Ierna, A.; Lombardo, G.M.; Mauromicale, G. Yield, nitrogen use efficiency and grain quality in durum wheat as affected by nitrogen fertilization under a Mediterranean environment. *Expl. Agric.* **2016**, *52*, 314–329. [CrossRef]

38. ICCP, Climatic Change 2012. *Managing the Risks of Extreme Events and Disasters to Advance Climate Change Adaptation*; A Special Report of Working Groups I and II of the Intergovernmental Panel on Climate Change; Field, C.B., V. Barros, T.F., Stocker, D., Qin, D.J., Dokken, K.L., Ebi, M.D., Mastrandrea, K.J., Mach, G.K., Plattner, S.K., Allen, M., Eds.; Cambridge University Press: Cambridge, UK, 2012.

39. Nuttall, J.G.; O'Leary, G.J.; Panozzo, J.F.; Walker, C.K.; Barlow, K.M.; Fitzgerald, G.J. Models of grain quality in wheat—A review. *Field Crop Res.* **2015**, *202*, 136–145. [CrossRef]

40. Cossani, C.M.; Slafer, G.A.; Savina, R. Nitrogen and water use efficiencies of wheat and barley under a Mediterranean environment in Catalonia. *Field Crops Res.* **2012**, *128*, 109–118. [CrossRef]

41. López-Bellido, L.; Mu˜noz-Romero, V.; Benítez-Vega, J.; Fernández-García, P.; Redondo, R.; López-Bellido, R.J. Wheat response to nitrogen splitting applied to a vertisols in different tillage systems and cropping rotations under typical Mediterranean climatic conditions. *Eur. J. Agron.* **2012**, *43*, 24–32. [CrossRef]

42. Triboï, E.; Martre, P.; Girousse, C.; Ravel, C.; Triboï-Blondel, A.M. Unravelling environmental and genetic relationships between grain yield and nitrogen concentration for wheat. *Eur. J. Agron.* **2006**, *25*, 108–118. [CrossRef]

43. Giuliani, M.M.; Giuzio, L.; De Caro, A.; Flagella, Z. Relationships between Nitrogen Utilization and Grain Technological Quality in Durum Wheat: II. grain yield and quality. *Agron J.* **2011**, *103*, 1668–1675. [CrossRef]

44. Li, Y.; Yin, Y.P.; Zhao, Q.; Wang, Z.L. Changes of glutenin subunits due to water-nitrogen interaction influence size and distribution of glutenin macropolymer particles and flour quality. *Crop Sci.* **2011**, *51*, 2809–2819. [CrossRef]

45. Malik, A.H.; Kuktaite, R.; Johansson, E. Combined effect of genetic and environmental factors on the accumulation of proteins in the wheat grain and their relationship to bread-making quality. *J. Cereal Sci.* **2013**, *57*, 170–174. [CrossRef]

46. Bouacha, O.D.; Rhazi, L.; Aussenac, T.; Rezgui, S.; Nouaigui, S. Molecular characterization of storage proteins for selected durum wheat varieties grown in different environments. *J. Cereal Sci.* **2015**, *61*, 97–104. [CrossRef]

47. Labuschagne, M.T.; Elago, O.; Koen, E. Influence of extreme temperatures during grain filling on protein fractions, and its relationship to some quality characteristics in bread, biscuit, and durum wheat. *Cereal Chem.* **2009**, *86*, 61–66. [CrossRef]

48. Corbellini, M.; Mazza, L.; Ciaffi, M.; Lafiandra, D.; Borghi, B. Effect of heat shock during grain filling on protein composition and technological quality of wheats. In *Wheat, Prospects for Global Improvement, Proceedings of the 5th International Wheat Conference, Ankara, Turkey, 10–14 June 1996*; Springer: Berlin/Heidelberg, Germany, 1998.

49. Wardlaw, I.F.; Blumenthal, C.S.; Larroqu, O.; Wrigley, C.W. Contrasting effects of chronic heat stress and heat shock on kernel weight and flour quality in wheat. *Funct. Plant Biol.* **2002**, *29*, 25–34. [CrossRef]

50. Rossmann, A.; Buchner, P.; Savill, G.P.; Hawkesford, M.J.; Scherf, K.A.; Mühling, K.H. Foliar N application at anthesis alters grain protein composition and enhances baking quality in winter wheat only under a low N fertiliser regimen. *Eur. J. Agron.* **2019**, *109*. [CrossRef]

51. Loussert, C.; Popineau, Y.; Mangavel, C. Protein bodies ontogeny and localization of prolamin components in the developing endosperm of wheat caryopses. *J. Cereal Sci.* **2008**, *47*, 445–456. [CrossRef]

52. Koga, S.; Aamot, H.U.; Uhlen, A.K.; Seehusen, T.; Veiseth-Kent, E.; Hofgaard, I.S.; Moldestad, A.; Böcker, U. Environmental factors associated with glutenin polymer assembly during grain maturation. *J. Cereal Sci.* **2020**, *91*. [CrossRef]

53. Shewry, P.R.; Toscano Underwood, C.; Wan, Y.; Lovegrove, A.; Bhandari, D.; Toole, G.; Mills, E.N.; Denyer, K.; Mitchell, R. Storage product synthesis and accumulation in developing grains of wheat. *J. Cereal Sci.* **2009**, *50*, 106–112. [CrossRef]

MDPI

St. Alban-Anlage 66

4052 Basel

Switzerland

Tel. +41 61 683 77 34

Fax +41 61 302 89 18

www.mdpi.com

Agronomy Editorial Office

E-mail: agronomy@mdpi.com

www.mdpi.com/journal/agronomy

9 783039 431021